T0299053

SOCIETY FOR THE STUDY OF HUMAN BIOLOGY
SYMPOSIUM SERIES: 26

Capacity for work in the tropics

PUBLISHED SYMPOSIA OF THE
SOCIETY FOR THE STUDY OF HUMAN BIOLOGY

1 The Scope of Physical Anthropology and its Place in Academic Studies   Ed. D. F. ROBERTS and J. S. WEINER
2 Natural Selection in Human Populations   Ed. D. F. ROBERTS and G. A. HARRISON
3 Human Growth   Ed. J. M. TANNER
4 Genetical Variation in Human Populations   Ed. G. A. HARRISON
5 Dental Anthropology   Ed. D. R. BROTHWELL
6 Teaching and Research in Human Biology   Ed. G. A. HARRISON
7 Human Body Composition, Approaches and Applications   Ed. J. BROZEK
8 Skeletal Biology of Earlier Human Populations   Ed. D. R. BROTHWELL
9 Human Ecology in the Tropics   Ed. J. P. GARLICK and R. W. J. KEAY
10 Biological Aspects of Demography   Ed. W. BRASS
11 Human Evolution   Ed. M. H. DAY
12 Genetic Variation in Britain   Ed. D. F. ROBERTS and E. SUNDERLAND
13 Human Variation and Natural Selection   Ed. D. F. ROBERTS (Penrose Memorial Volume reprint)
14 Chromosome Variation in Human Evolution   Ed. A. J. BOYCE
15 Biology of Human Foetal Growth   Ed. D. F. ROBERTS
16 Human Ecology in the Tropics   Ed. J. P. GARLICK and R. W. J. KEAY
17 Physiological Variation and its Genetic Basis   Ed. J. S. WEINER
18 Human Behaviour and Adaptation   Ed. N. G. BLURTON JONES and V. REYNOLDS
19 Demographic Patterns in Developed Societies   Ed. R. W. HIORNS
20 Disease and Urbanisation   Ed. E. J. CLEGG and J. P. GARLICK
21 Aspects of Human Evolution   Ed. C. B. STRINGER
22 Energy and Effort   Ed. G. A. HARRISON
23 Migration and Mobility   Ed. A. J. BOYCE
24 Sexual Dimorphism   Ed. F. NEWCOMBE et al.
25 The Biology of Human Ageing   Ed. A. H. BITTLES and K. J. COLLINS
26 Capacity for Work in the Tropics   Ed. K. J. COLLINS and D. F. ROBERTS
27 Genetic Variation and its Maintenance   Ed. D. F. ROBERTS and G. F. DE STEFANO

Numbers 1–9 were published by Pergamon Press, Headington Hill Hall, Headington, Oxford OX3 0BY. Numbers 10–24 were published by Taylor & Francis Ltd, 10–14 Macklin Street, London WC2B 5NF. Further details and prices of back-list numbers are available from the Secretary of the Society for the Study of Human Biology.

Capacity for work in the tropics

Edited by
K. J. COLLINS
Medical Research Council, London School of Hygiene and Tropical Medicine, London
D. F. ROBERTS
Department of Human Genetics, University of Newcastle upon Tyne

CAMBRIDGE UNIVERSITY PRESS
Cambridge
New York   New Rochelle
Melbourne   Sydney

CAMBRIDGE
UNIVERSITY PRESS

University Printing House, Cambridge CB2 8BS, United Kingdom

Cambridge University Press is part of the University of Cambridge.

It furthers the University's mission by disseminating knowledge in the pursuit of education, learning and research at the highest international levels of excellence.

www.cambridge.org
Information on this title: www.cambridge.org/9780521309356

First published 1988

A catalogue record for this publication is available from the British Library

ISBN 978-0-521-30935-6 Hardback
ISBN 978-0-521-11863-7 Paperback

CONTENTS

Page

Preface   vii

Measurement of Working Capacity in Populations

R. J. Shephard: 1
Work capacity: methodology in a tropical environment

J. M. Patrick: 31
Ventilatory capacity in tropical populations: constitutional
and environmental influences

J. E. Cotes: 51
Is measurement of aerobic capacity a realistic objective?

Functional Consequences of Malnutrition

R. Martorell & G. Arroyave: 57
Malnutrition, work output and energy needs

J.V.G.A. Durnin & S. Drummond: 77
The role of working women in a rural environment when
nutrition is marginally adequate: problems of assessment

J. D. Haas, D. A. Tufts, J. L. Beard, R. C. Roach & H. Spielvogel: 85
Defining anaemia and its effect on physical work capacity
at high altitudes in the Bolivian Andes

G. B. Spurr: 107
Marginal malnutrition in childhood: implications for adult
work capacity and productivity

A. Ferro-Luzzi: 141
Marginal energy malnutrition: some speculations on primary
energy sparing mechanisms

Page

Growth, Stature and Muscular Efficiency

J. Ghesquiere & C. D'Hulst   165
    Growth, stature and fitness of children in tropical areas

Ethnic and Socio-Cultural Differences in Working Capacity

D. F. Roberts: 181
    Genetics of working capacity

J. Huizinga: 193
    Working capacity in different African groups

S. Samueloff: 205
    Environmental, genetic and leg mass influences on energy
    expenditure

C. M. Beall & M. C. Goldstein: 215
    Sociocultural influences on the working capacity of elderly
    Nepali men

Energy Expenditure and Endemic Disease

E. Bengtsson, P. O. Pehrson, A. Bjorkman, J. Brohult, L. Jorfeldt,   227
P. Lundbergh, L. Rombo, M. Willcox & A. Hanson:
    Malaria: working capacity in a holo- and mesoendemic
    region of Liberia

K. J. Collins, T. A. Abdel-Rahaman & M. A. Awad El Karim: 235
    Schistosomiasis: field studies of energy expenditure in
    agricultural workers in the Sudan

Research Models in Tropical Ecosystems

R. Brooke Thomas, T. L. Leatherman, J. W. Carey & J. D. Haas: 249
    Biosocial consequences of illness among small scale
    farmers: a research design

D. J. Bradley, L. Rahmathullah & R. Narayan: 277
    The tea plantation as a research ecosystem

Index   289

PREFACE

Following the decision of the International Union of Biological Sciences in 1982 to embark on a new research programme, "The Decade of the Tropics", the importance of including some themes relating to human biology was immediately apparent. There are so many human problems that are outstanding among tropical populations, and for a significant contribution to be made by the programme it would be necessary to concentrate effort on a few of the more important topics. Canvassing human biologists throughout the world on those investigations currently in progress and those thought important to develop, brought a list of some fifty projects. From these, two areas emerged as requiring urgent attention: one concerned the study of genetic variability, whilst the other concerned variations in working capacity and factors affecting it in tropical populations. To provide a critical review of present knowledge on work performance in the tropics, which would help to identify areas for further research, a meeting of scientists of high international standing was held at the CIBA Foundation in London on December 13th and 14th, 1984. The papers discussed then are published in the present volume.

For the purpose of the scientific meeting, the topics were arranged under six general headings: measurement of working capacity in populations; functional consequences of malnutrition; growth, stature and muscular efficiency; ethnic differences in working capacity; energy expenditure and endemic disease; and research models in tropical ecosystems. The present volume retains this framework, and all the studies reported contain new research carried out during the last ten years.

In many respects, the proposed research programme would take up the mantle of the International Biological Programme (1964-1974), the Human Adaptability Section of which contributed so much to understanding the interrelationships between human populations and their environment. It has since become clear from the many promising lines of research on the theme of work performance in the tropics, some of which are reported here, that much of the recent work has complemented the initial multidisciplinary projects started by the I.B.P. ˙In general, it was felt that the aim of further research effort should be to document the biological effects of changes occurring now in the tropical biome, both on the working capacities and the requirements of the human communities themselves, and how these in turn affect the ecological systems of which they form part.   Though not a fully comprehensive account of current work in the tropics, this volume brings together many of the major research topics which, it is hoped, will form a basis on which to launch the new programme.

A special word of appreciation is extended to Professors P. T. Baker, H. M. Gilles, G. A. Harrison, J. M. Tanner and J. C. Waterlow, who assisted as Chairmen and in discussion during the meeting.

K. J. COLLINS
Medical Research Council
London School of Hygiene and Tropical Medicine,
U.K.

D. F. ROBERTS
Department of Human Genetics
University of Newcastle upon Tyne
U.K.

MEASUREMENT OF
WORKING CAPACITY IN POPULATIONS

WORK CAPACITY:
METHODOLOGY IN A TROPICAL ENVIRONMENT

R. J. SHEPHARD

School of Physical & Health Education and Department of Preventive Medicine and Biostatistics, Faculty of Medicine, University of Toronto, Toronto, Ontario, Canada

INTRODUCTION

The tropical habitat currently faces strong pressures. In many parts of Africa, the current growth rate for human populations is 4.5-5.0% per annum, compared with the world average of 1.7%. While there remains some disagreement on minimal nutritional requirements (Rivers & Payne, 1982; Weymes, 1982), the expanding deserts of the sub-Saharan poverty belt threaten starvation to many peoples of this region. At the same time, large parts of the African continent have very few inhabitants. In the "Decade of the Tropics" much thus depends on human ability to exploit the African habitat, producing a surplus of food that can nourish rapidly expanding cities.

One important determinant of the ability to colonize any harsh environment is the individual's physiological working capacity (Harrison & Walsh, 1974; Shephard, 1978), the "ability to perform muscular work satisfactorily" (Andersen et al, 1971). In this review, particular attention will be paid to developments since completion of the International Biological Programme (I.B.P., Worthington, 1978). The changing nature of work capacity will be discussed in a tropical context, along with issues such as sampling bias, technical problems of measurement in a hot environment, and procedures for allocating variance between constitution and environment. The effect of changes in body size, nutrition and health will also be considered briefly.

THE CHANGING NATURE OF WORK CAPACITY

The physiological characteristics appropriate to successful colonization of a given habitat depend on (i) population pressures, (ii) the amount of physical work necessary for survival, and (iii) available technology (Breymeyer & Van

Dyne, 1979).   Each of these factors is currently changing.   An ever-increasing proportion of tropical inhabitants are living in large cities.   In a few countries, such as Singapore, high technology has made rapid strides, although 80% of the population still work in "cottage" type situations, building components for the automated factories.   Elsewhere, the movement is from primitive agriculture to labour-intensive industry.   Thus in American Samoa (Greksa & Baker, 1982), 55% of the population now engage in paid manual labour, at an energy cost of 12-16 kJ/min;   8% are still agriculturalists, spending 30-32 kJ/min;   and the remaining 37% have become office workers.   In countries where conditions are less "advanced", the gross domestic product remains very small, increasing in real value by less than 1% per annum, while limited education precludes the drastic infusion of new technology.   The burden of feeding a growing population in such regions rests not on the export-earnings of the industrial workers but on the physical efforts of rural peasants using at best intermediate levels of agricultural technology (Brandt, 1980).   Attempts to improve productivity raise important social and economic questions (Collins, 1982).   Is the optimal approach to enhance the work capacity of the peasant, through better nutrition and the control of disease (Davies, 1974), or should investment be concentrated on the methodology of farming (improvement of crops, livestock, and tools; Andreano, 1976; Collins, 1982)?   Economic cost-benefit analysis is complicated by the fact that the harvesting of cash crops generally provides short-term employment, while improvements of nutrition and health are likely to have a more long-term impact upon work capacity (Bliss & Stern, 1982).

Tropical food-producers do not face the extreme climatic changes encountered in the circumpolar regions, and more specialization of economic activity is thus possible.   Nevertheless, there is no single "tropical environment" to which biological adaptation can be made.   Indeed, settlements are often located at the junction of two or more ecosystems in order to provide a variety of food (Harris, 1982).   In recent years, traditional work requirements have been further complicated by interactions with the economy of "white" migrants. Some tropical peoples still live on small islands where strength in paddling and skill in various forms of spear fishing are major assets.   Others, in the dusty sub-Saharan region, use simple hand tools to grow millet, sorghum and tubers (Richards, 1982).   Swampy flat land or rugged hillsides may also be cultivated (Ferro-Luzzi, 1982).   A few of the tropical peoples still run after game or walk long distances searching for nuts and berries (Harris, 1982).   It is most unlikely that any one set of physiological characteristics would confer a selective advantage in exploiting all of these varied techniques of survival at both wet and dry seasons.   A more general evaluation of human work capacity is required.

THE I.B.P. CONCEPT OF WORK CAPACITY AND HUMAN ADAPTABILITY

The I.B.P. evaluation of working capacity (Shephard, 1978) included data on standing height, body mass, body fat, muscle strength, lung volumes, anaerobic capacity and power. Nevertheless, it was argued that the main determinant of the ability to perform most types of daily work was the maximum oxygen intake ($\dot{V}_{O_2}$ max; Shephard, 1977). This was sometimes expressed in absolute terms (l/min), but where the daily activity required a displacement of body mass, relative units (ml/kg/min) were more appropriate. Expression of $\dot{V}_{O_2}$ max as a ratio to lean body mass was shown not to increase the quality of the information, and indeed in the context of comparing people of widely differing builds, it could be frankly misleading (Gitin et al, 1974).

Much agricultural activity remains quite hard endurance work, whether pursued at a primitive or an intermediate level of technology (Shephard, 1985), and in such situations a large maximum oxygen intake might be thought advantageous. However, whether searching for game, foraging for nuts, or attempting to grow vegetables in unpromising soil, the yield of food depends more on technique than on brute force. Moreover, if the population is self-employed, a relative abundance of food (or poor prices for cash crops) may hold the average person's production far below the physiological potential (Lee, 1969; Ekblom & Gjessing, 1968; Harris, 1982). In New Guinea, a combination of hunting and the tending of cash crops occupy the Kaul for 3.0-3.5% and the Lufa for 6.0-7.5% of the day (Norgan & Ferro-Luzzi, 1978). Nevertheless, the apparently light occupational demands of such a society are supplemented by a need for domestic labour that is avoided in a more highly developed economy (Hawkes & O'Connell, 1981). Thus Clark and Haswell (1964) observed that African and Indian peasants not only spent 17-34 hours working in the field each week, but also carried out such activities as building and repairing houses, making shoes and furniture, spinning, sewing, and grinding corn.

A further difficulty in the interpretation of data is that in some regions any adaptations to a high environmental temperature become confounded with physical demands imposed by the cultivation of a hilly terrain (Patrick & Cotes, 1971, 1978).

Collation of I.B.P. results for various tropical populations (Table 1) provided no evidence for the emergence of physiologically well-adapted variants, either within specific tropics regions, or relative to other populations that have exploited a temperate or a cold habitat. The body mass of the tropical societies was generally very low relative to sedentary "white" standards, and with a few

Table 1:   Physiological variables affecting working capacity of labourers in a
tropical environment.   Range of data reported during IBP studies
(from Shephard, 1978).   All relate to males, unless otherwise
specified.

Variable	Range	Location of low and high extremes
Excess body mass (Kg)	-13.0 to 2.5	Ethiopia, Yemenite-Israelis
Skinfold thickness (mm)	4.5 to 10.6	Kalahari bushmen to Trinidadian East Indians
Muscle strength:		
handgrip (N)	339 to 417	Ethiopia, Warao, Venezuela
knee extension force (N)	288 to 364	Ethiopia, Warao, Venezuela
Anaerobic power (ml/kg/min)	125*	Values for East Africans only.
capacity (ml/kg)	-	No data
Aerobic power (ml/kg/min)		
Male	37.6 to 57.2	Rural Bantu, "active" Tanzanians
Female	27.3 to 41.0	Jamaicans, East African-Dorobo

*This calculation has followed Margaria (1966) in assuming a 25% mechanical
efficiency for the staircase sprint. In fact, resynthesis of phosphagen does not
occur within the period of observation, so that the equivalent oxygen consumption may be substantially larger than postulated by Di Prampero (1972).

exceptions, skinfold thicknesses averaged less than 7 mm.   Muscle strength and
anaerobic power were only about 75% of the figures found in Europeans, but
partly because of a low body mass, most tropical populations had a relative
anaerobic power at least as large as their European counterparts.

DEVELOPMENTS IN METHODOLOGY

There have been several technical developments in the measurement of
work capacity since completion of the I.B.P. project.

Body Mass

Young women and the elderly of both sexes have increasingly demanded
access to heavy employment, particularly in more sophisticated societies.
Physiologists have recognised that lifting is an important component of the
ability to perform heavy work and that body mass provides a simple indication of
the individual's potential in this regard (Nottrodt & Celentano, 1984).   In part
because of a negative correlation between body mass and cigarette smoking

Table 2: Discrepancy of body mass between male Indian athletes (I) and competitors at the Montreal Olympic Games (O). Based in part on data accumulated by Sodhi and Sidhu (1984).

Event	Mass ($\frac{I - O}{O}$%)	Height ($\frac{I - O}{O}$%)	Event	Mass ($\frac{I - O}{O}$%)	Height ($\frac{I - O}{O}$%)
Jumping	-18.5	-9.3	400 m	-16.0	-11.4
Pole vault	-11.5	-5.8	800 m, 1500 m	-17.0	-11.6
Discus	-10.5	-9.6	Marathon	-9.3	-11.2
Javelin	-23.8	-11.8	110 m hurdles	-16.7	-11.0
100 m, 200 m track	-15.0	-11.1	400 m hurdles	-9.6	-6.0

(Garrison et al, 1983), the Metropolitan Life Assurance Company (1983) has proposed an upward revision of the actuarial "ideal" body mass. However, the concept that low-income tropical societies have a low body mass relative to their height remains unchallenged. Thus Indian athletes still show a deficit of 5.6 to 22.9 kg (9.6 to 23.8%) relative to "white" competitors, only about a half of this difference being attributable to their shorter stature (Sodhi & Sidhu, 1984; Table 2). Nigerian basketball players are even smaller than their Indian counterparts (Mathur, 1982). It remains unclear whether any of this mass deficit is constitutional, and whether the life expectancy of the tropical groups would be improved by a substantial weight gain.

Body fat

Controversy continues over methods of estimating body fat. There have been renewed suggestions that at least in a well-nourished population, the body mass index (mass/height^2) provides the best simple field index of fatness (Baecke, 1982). One important practical lesson from the I.B.P. was that this index loses its validity if muscle mass is increased by heavy physical work, or is decreased by chronic malnutrition. The interpretation of skinfold data may also be complicated in poorly nourished populations; apparently, "essential" trunk fat is conserved, while fat is lost from the arms (Bogin & MacVean, 1981; Johnston et al, 1984). There are many additional problems in trying to convert skinfold readings to estimates of body fat, including possible differences in the distribution of body fat between superficial and deeper body depots, and a two-fold variation in fat estimates according to the choice of prediction formulae (Shephard, 1978; Lohman, 1981; Cureton, 1983). In the tropics, the "normal" variation in density of the several body compartments is often enhanced by a lack of bone mineral and increases of tissue water content secondary to malnutrition. Jones & Corlett

(1980) demonstrated a correlation between total limb width and the mineral content of the femur among Indians; the fat-free density of the Ghurkas was 0.003 above the standard value, leading to a 15% error in the calculated fat-free mass.

Muscle strength

Interest in muscle strength as a factor limiting the employment of women and elderly men has led to the development of an incremental lifting task (Nottrodt & Celentano, 1984); however, to date, this and other tests of muscle strength have yielded little more information about work capacity than that provided by a determination of body mass. Many exercise physiology laboratories now supplement "static" measurements of muscle strength with isokinetic or isotonic data, recorded from Cybex and Ariel dynamometers respectively. The equipment is expensive, not readily portable, and does not yet appear to have been applied to tropical populations. Fortunately, there seems to be a fairly close correlation between isokinetic data and the corresponding isometric readings (Kofsky, Davis & Shephard, 1985); on the other hand, the isokinetic data bears little relationship to the scores for tests of explosive strength (such as a vertical jump, Genuerio & Dolgener, 1980). In theory, lean body mass is an attractive indicator of muscularity, since little cooperation is required from the subject (Forbes, 1974). Low values confirm poor muscular development (Miller et al, 1972), but most available methods of estimating lean mass suffers from the problems of varying body composition already discussed.

The weak muscles of an undernourished, tropical resident are plainly a disadvantage in either agricultural or industrial tasks that call for substantial strength, but because of the small amount of body fat, the overall handicap may be less than force measurements suggest (Petrofsky & Phillips, 1981). Difficulty in perfusing weak muscles may also limit heavy aerobic effort (Kay & Shephard, 1969).

Muscle biopsy

A further, important development in the evaluation of muscle function has been a perfection of needle biopsy techniques. New information has been obtained on the distribution of muscle fibre-types, on local stores of water, minerals and metabolites, and on muscle enzyme activities. This approach should be of particular interest in monitoring human adaptability to the tropical habitat, since there is solid evidence that immediate environmental conditions (including hard physical work) cannot convert a slow-twitch fibre to a fast-twitch fibre, or vice-versa (Saltin et al, 1977).

Anaerobic power and capacity

I.B.P. estimates of anaerobic power and capacity were based on scores for a staircase sprint (Margaria, 1966) and determinations of blood lactate (DiPrampero, 1971) respectively. East Africans, in common with many "primitive" groups, fared poorly on these tests. However, it was unclear whether the problem was anatomical (differences in geometry of the limb muscles), physiological (deficiency of muscle mass or of phosphagen due to malnutrition) or experimental (including difficulties of communication with the subjects, their lack of experience with the required test procedures, and an absence of competitive spirit in the societies under evaluation). It is tempting to attribute much of the functional deficit to a nutritionally based limitation of muscular development (Malina, 1984). However, malnutrition also begets inactivity and thus loss of physical condition, while the power output that can be demonstrated in a staircase sprint varies to some extent with the mass that is to be lifted (Caiozzo & Kyle, 1980). Alternative, cycle-ergometer field tests of anaerobic capacity and power are now available (Bar Or, 1983), although it has yet to be demonstrated that a tropical villager finds the fast pedalling of a cycle ergometer any easier than the task of sprinting up a flight of stairs. Indeed, unless ergometer loadings are adjusted for the small leg volume of tropical residents, it seems likely that misleading answers will be obtained (Lavoie et al, 1984).

Some investigators continue to use performance tests as measures of anaerobic work capacity. Procedures have included a 50 metre run (Ruskin, 1978; U.S. Sports Academy, 1983), a 256 metre run (Thompson, 1981), a medicine ball throw and a vertical jump (Ghesquiere & Eckles, 1984). The last authors plotted their data against body mass, in an attempt to overcome the well-recognized effects of body size upon performance test scores. The results obtained in a tropical milieu are also depressed by lack of test experience, a hot environment, poor track conditions and the absence of a sense of competition (Strydom, 1978).

Aerobic power

Much additional investigation has substantiated the I.B.P. concept that the maximum oxygen intake measured by uphill treadmill running provides a generalisable measure of aerobic power that cannot be materially exceeded (Thys et al, 1979; Shephard, 1982, 1984). It has been repeatedly confirmed that maxima are at least 3-4% smaller when stepping, and 7-8% smaller during cycle ergometry. However, much of the deficit encountered in a cyclist can be corrected by the use of a racing design of cycle, with optimal gear ratios and drop handlebars (Faria et al, 1978), rat-trap pedals (King et al, 1982) and standing on the pedals (Kelly et al, 1980).   Irrespective of the mode of exercise that is adopted, the quality control of data is of great importance (Jones & Kane, 1979).   Under field conditions, the validity of data is best assured by the regular testing of a panel of volunteers with known characteristics.

In Canada, a step test developed for the field-testing of aerobic power has found extensive application in population surveys and self-administered fitness evaluation (Bailey et al, 1976;   Fitzgerald et al, 1980).   The concept of using music to set the rhythm of stepping evolved during our studies of the Inuit, but with an appropriate adjustment of instrumentation, music could also provide a useful method of teaching and motivating tropical populations during step-testing.   One point to check for a tropical population is the mechanical efficiency of step-climbing;   possibly as a part of their heat acclimatisation, indigenous populations are sometimes more efficient than those from cooler regions (Wyndham, 1966).

Other designs of step test (Wyndham & Sluis Cremer, 1968; Robertshaw et al, 1984) continue to find industrial field application both in South Africa and the U.K.   All-out cycle ergometer tests have also been extrapolated to the oxygen consumption at the presumed maximal heart rate ($\dot{V}O_{2,195}$:   Collins, 1982).   All-out runs over distances of 402-8050 metres have not had thorough evaluation as a simple method of predicting PWC170 (Table 3).   There is some concern regarding the safety of prolonged, unsupervised running in older populations, and in general, correlations with the directly measured maximum oxygen intake are only moderate.   Scores depend greatly upon motivation, knowledge of pacing, and environmental conditions (Shephard, 1982), with the largest coefficients of correlation being reported over distances of 2000-3000 metres.

Anaerobic threshold

One final aspect of work performance that has attracted much attention over the past decade has been the relative intensity of effort at which significant amounts of lactate accumulate in the blood stream - the so-called "anaerobic threshold".   It has been suggested that well-trained and well-muscled individuals can operate at a large fraction of their maximum oxygen intake without a significant build-up of lactate, and this apparently gives them an advantage in tasks calling for sustained endurance (Ready & Quinney, 1982; Powers et al, 1983).   In general, determinations of both the aerobic threshold (where Type I muscle fibres are heavily recruited and an imbalance develops

Table 3: Coefficients of correlation between directly measured maximum oxygen intake and running speed (based on reports tabulated by Shephard, 1982).

Distance	Coefficient of correlation	Number of reports
402 m	0.22-0.31	2
549 m	0.27-0.67	6
805 m	0.04-0.73	6
1610 m	0.25-0.79	5
9 min	0.71-0.81	2
12 min	0.28-0.90	11
3220 m	0.47-0.86	3
4830 m	0.43	1
8050 m	0.38	1

between pyruvate formation and oxidation), and the anaerobic threshold (above which large amounts of lactate accumulate in the blood stream) have required access to a laboratory with facilities for on-line determinations of gas exchange. However, the second and more clearly identified breakpoint in the gas exchange curve can be estimated from its correlation with the time to exhaustion on a simple cycle ergometer test (Jacobs et al, 1983), with the perception of "somewhat hard" work on the Borg scale of perceived exertion (Purvis & Cureton, 1981), with absolute heart rate (Parkhouse et al, 1982), and with a non-linearity in the relationship between heart rate and work rate (Conconi et al, 1982).

Collins (1982) has illustrated some of the difficulties encountered when translating these physiological and psycho-physical measurements into estimates of productive work. In a commercial situation such as a sugar plantation, bonuses and experience of cane-cutting are generally more important variables than any estimate of physiological work capacity.

MEASURING WORK CAPACITY IN A TROPICAL ENVIRONMENT

Both techniques of testing and the interpretation of data may need modification in the tropical environment.

Sampling problems

Sampling presents many practical problems. Identification of names and ages becomes imprecise when these have not been recorded by the government (Sinnett & Whyte, 1973; Davies et al, 1974). In some studies, sampling appears to have been fairly complete; for example, Goldrick et al (1970) examined 95% of the 1,500 member Enga clan in the New Guinea highlands. Other reports give no indication of the proportion of the total population that has been sampled.

By analogy with the circumpolar data, incomplete sampling usually leads to over-representation of the fit, and under-representation of the diseased (Shephard, 1978). This has serious implications for population estimates of working capacity in countries where conditions such as malnutrition and anaemia (Fleming, 1977), and diseases such as malaria, schistosomiasis, hookworm, filariasis and amoebic infections are endemic (Davies & Van Haaren, 1974; Huizinga & Hooijen Bosma, 1980; Weymes, 1982; Collins, 1982; Ghesquiere & Eekels, 1984; Van Ee, 1985).

Other, more general problems of sampling can also arise in the tropics. Volunteers display their need of peer approval by out-performing randomly selected individuals and exaggerating their normal performance both in questionnaire responses and in field measurements of energy expenditure (Arnett & Rikli, 1981). Urban samples are further biased by a limited recruitment of smokers and alcoholics.

Body build

Despite the supposed advantages of heat loss conferred by a linear body form, a wide range of statures is encountered in the tropics, from the pygmy Twa (average height 144 cm) to the inhabitant of the arid sub-Saharan region (average height, 171 cm; Hiernaux, 1966; Austin et al, 1979). This immediately raises the vexing issue of proportionality in measurements of working capacity. For example, if maximum oxygen intake varied as height$^{3.0}$, a Saharan man would have a 67% advantage of oxygen transport over the Twa. However, if the exponent were $H^{2.0}$, the advantage would drop to 41%. Details of proportionality theory (Von Döbeln, 1966; Ross et al, 1978) are unfortunately not borne out by experimental data collected over the period of normal growth (Table 4), and it seems rather improbable that such calculations can adjust for differences of size between adult populations.

A further complication is that a secular trend associated with greater hybridisation and changes of nutrition has increased the body size of many groups over recent years; for example, the Twa are now 10-16 cm taller than the classical figure (Ghesquiere, 1971; Austin et al, 1979). Socioeconomic and/or nutritional gradients of stature have been described in India (Malhotra, 1966), Tunis (Pařízková et al, 1972), Samoa (Greksa & Baker, 1982), and Kinshasa (Ghesquiere & Eekles, 1984). The authors of the last study commented that when the results of several field performance tests were expressed relative to body mass (approximately an $H^{3.0}$ correction), the poorer segment of the

Table 4: Influence of stature upon work capacity measurements: comparison between theoretical relationship (Von Döbeln, 1966) and observed relationship in children aged 6 to 12 years (Shephard et al, 1980).

Variable	Theoretical relationship	Observed relationship
Body mass	$H^{3.0}$	
Static muscle force	$H^{2.0}$	$H^{3.2}$
Aerobic power	$H^{2.0}$	$H^{2.78}$
PWC_{170}	$H^{2.0}$	$H^{3.37}$
Vital capacity	$H^{3.0}$	$H^{2.68}$
Running speed	$H^{0.0}$	$H^{0.81}$

community showed no disadvantage of fitness despite its smaller size. However, others have found that poor nutrition reduces aerobic power (Viteri, 1971; Spurr et al, 1974) and the endurance of submaximal work (Satyanarayana et al, 1978) even after allowance for dimensional differences. Spurr et al (1974) suggested that poor performance could be corrected by nutritional rehabilitation, but Viteri (1971) found that after three years of dietary supplementation, Guatemalan paeons showed only an insignificant increase in the work performed to exhaustion, with no augmentation of their maximum oxygen intake.

There have been suggestions of an altered distribution of body fat in poorly nourished subjects (Eveleth, 1979). The standard prediction equations of Durnin and Womersley (1974) apparently yield appropriate estimates of body fat for Indian men, Chilean men, Taiwanese, New Guinean women and young men. However, predictions seem unsatisfactory in Indian women (Satwanti et al, 1977) and older men in New Guinea (Norris, 1982). The problem has practical importance, since acculturation to a sedentary "western" lifestyle often leads to an increase of relative body mass and of body fat, with a decrease of relative aerobic power (Prior & Evans, 1970; Mann et al, 1955; Greksa & Baker, 1982). In Africa, rural workers still remain thin (Table 5), and perhaps for this reason, urbanisation has been held to improve their working capacity (Vorster, 1977).

Hydration

Dehydration can cause changes in both body composition and maximum oxygen intake (Wyndham, 1967). A sustained bout of exercise in a warm environment induces a fluid loss of several litres, even if replacement fluids are available. Such disturbances are compounded by a loss of a further 1.5 kg of water if glycogen reserves are depleted by either starvation or prolonged

Table 5:   Some skinfold estimates of percent body fat in African populations
(based in part on data collected by Norgan, 1982).

Men		Women	
Population	Estimated percent fat	Population	Estimated percent fat
Masai - tribal	6	Upper Volta	19
- non-tribal	6	(farmers)	
Samburu - tribal	9	Tanzanians (urban)	26
- non-tribal	12		
Tanzanians (urban)	12		
Nigerians (soldiers)	12		

exercise.   In contrast, a lack of plasma protein can boost the water content of
the tissues through oedema formation.   Thus, care must be taken to ensure
normal hydration when assessing either body mass or body composition.   One
example of the problem that can arise from a disturbed fluid balance seems a
report that undernourished Columbian men have 15-20% body fat (Barac-Nieto et
al, 1978).

Muscle strength

Recent observations from Bahrein (U.S. Sports Academy, 1983) confirm
the view that populations indigenous to hot areas have much lower handgrip
readings than those living in "western" socities (Table 6).   The height
differential is about 10 cm, so that even on the assumption that muscle strength
is proportional to $H^{3.0}$ we could explain only 16-17% of the discsrepancy from
Western data.   Poor nutrition seems unlikely among the relatively wealthy
population of Bahrein; other factors contributing to the low scores include poor
motivation, and a lack of habitual activity in extreme heat.

Maximum oxygen intake

Short-term exposure to heat has little influence upon the directly
measured maximal oxygen intake, but prolonged physical activity in a hot
climate reduces central blood volume and thus oxygen transport (Pirnay et al,
1970; Shephard, 1983).   Interpretation of $\dot{V}O_2$ max scores for tropical populations can again be complicated by issues of body size and nutritional status.
Most results have been reported per kg of body mass (approximately equivalent
to the assumption that scores vary as $H^{3.0}$) (Table 7).   Even if muscle mass is
poorly developed, oxygen transport per kg of body mass provides a practical

Table 6: Handgrip force (Newtons) as percentage of recent data for young Canadian men (524N, Canada Fitness Survey, 1983)

Population	Handgrip force (%)
Addis Ababa, 1969 (workers)	64.8
(airforce cadets)	77.5
Warao (Venezuela), 1971	79.6
Bahrein, 1983	71.6

empirical index of available power. Another alternative is to relate maximum oxygen intake to simple anthropometric estimates of leg volume (muscle + bone, Jones & Pearson, 1970; Davies & Van Haaren, 1973). Likewise, if the haemoglobin level of an anaemic subject is corrected by iron therapy, maximum oxygen intake may increase without a concomitant increase of leg volume (Davies, 1974). Finally, a high capacity for heavy physical work is associated with a large cycle ergometer maximum oxygen intake relative to leg volume (Davies, 1974).

As already noted with muscle strength, extreme heat may also impair oxygen transport by discouraging all physical activity. Further detraining of the legs may occur in swampy areas because the subject works while submerged to the waist (Collins, 1982).

Effects from a hot laboratory

Hot weather inevitably decreases motivation in all tests that require cooperation; this is particularly a problem in field tests for the estimation of aerobic power. Over the longer running distances, speeds are further decreased because a build-up of body heat diverts blood flow from the working muscles to the skin (Lloyd, 1966). The only published comments are by Larson (1974) "extremes of temperature should be avoided" and Strydom (1978) "the majority of tests are not seriously affected by the heat".

If maximum oxygen intake is predicted from the heart rate response to sub-maximal exercise, errors can arise from a heat-induced non-linearity of the heart rate/oxygen consumption line (Davies et al, 1971) and from a thermal tachycardia that exaggerates the normal heart rate response to effort. The latter is first observed at wet bulb temperatures above 25°C (Strydom, 1978). Unfortunately, tropical field measurements must sometimes be made at temper-

Table 7: Aerobic power of selected tropical populations (farmers, peasants and agricultural workers, based in part on data accumulated by Shephard, 1978; Norgan & Ferro-Luzzi, 1978; and Andersen et al, 1980).

Population	Maximum oxygen intake $(ml.kg^{-1}min^{-1})$	
	Male	Female
Bantu mine recruits	45	–
Columbian sugar cane cutters	45	
Dorobo, Turkana (E. Africa)	45	41
Easter Islanders	42	31
East African sugar cane cutters	48	
Ethiopians	35, 40	
Jamaican Hill Farmers	47	27
Kahalari Bushmen	47	
Kaul (New Guinea)	53	
Kurds	44	29
Malaysians	53	35
	(Temiars)	(Chinese Indians & Malaysian)
Nepalese	35	
Samoans	36–41	
Tanzanians – Active	57	
Inactive	47	40
Masai	54	
Trinidad – Negroes	38	
East Indians	38	
Warao (Venezuela)	45, 51	
Yemenites 47	35	
Yoruba (Nigeria)	49	32
Zairians –	40	25
Hoto	43	
Twa	48, 40	
Ntomba	44	

atures as high as 25-30°C, with a high associated relative humidity (Thinkaren et al, 1975; Duncan & Scammell, 1977; Greksa & Baker, 1982). One simple expedient is to play a fan upon the exercising subject, which reduces the effective temperature by about 2°C. Alternatively, a test may be used which depends more on physical work performance and less on heart rate. Quinney et al (1983) reported that the Canadian Home Fitness Test (which considers both rate of stepping and heart rate) is unaffected by environmental temperatures over the range 21-30°C (at a relative humidity of 70% throughout).

Hot conditions may also lower the anaerobic threshold. A falling systolic pressure leads to difficulty in perfusing the active muscles (Williams et al, 1962), while a reduction of visceral flow curbs the hepatic metabolism of lactate (Rowell et al, 1968). Finally, the oxygen cost of a given work task may be increased in the heat (Rowell, 1974). Thus, Negev Israelis expend 4-11% more energy on their various daily activities in the summer months than under thermoneutral conditions (Gold et al, 1969). In the early stages of exercise, a high ambient temperature gives some advantage from a lower muscle viscosity, but as activity continues, high temperatures augment oxygen costs through (i) a decreased efficiency in the transfer of phosphorylated energy within the active muscles, (ii) an increase of oxygen consumption by "inactive" tissues, (iii) poor mechanical efficiency secondary to fatigue, and (iv) the added metabolic cost of increases in cardiac output, ventilation and sweating (MacDougall et al, 1974; Shephard, 1982).

MEASURING CONSTITUTIONAL ADAPTATIONS

I.B.P. attempts to demonstrate the influence of constitution by comparing adaptations to various extreme habitats were largely unsuccessful (Shephard, 1978). Likewise, comparisons of different ethnic groups living in a similar environment were thwarted by substantial differences of diet, physical activity, body size and socio-economic status between the groups compared (Davies et al, 1972; Glick & Schvartz, 1974).

Formal twin studies (Weber et al, 1976) also proved disappointing, with heritability estimates varying wildly from one study to another (Shephard, 1978). The basic problem with this last type of analysis has been the unwarranted assumption of similarity in environmental and methodological variances for monozygotic twins, dizygotic twins, and members of the general population (Cotes, 1974; Shephard, 1978).

The calculation of parent-parent, parent-offspring and full sib-sib correlations has allowed the variance of work capacity data to be partitioned into four components - the environmental, the additive, dominant and interactive portions of genetic variance (Cavalli-Sforza & Bodmer, 1971). However, the theoretical basis of this approach seems weakened by differences in maternal-sibling and paternal-sibling comparisons (Bouchard, 1978; Montoye & Gayle, 1978; Bouchard et al, 1981). Presumably, the respective parents are influencing their offspring to differing extents through non-genetic mechanisms, including the foetal, neonatal and childhood environments.

MEASURING PATTERNS OF HABITUAL ACTIVITY

The examination of habitual activity patterns is an important foil to the measurement of work capacity.   It indicates whether the observed score matches the potential that would be realised with an adequate level of daily exercise.   In line with this concept, Cotes & Davies (1969) have suggested that fitness should be regarded as the proportion of the genetic endowment which has been realised.

Cyclic variations

The I.B.P. protocol stressed the need to sample typical activities of both summer and winter, including weekdays and weekend days in the analysis.   In "primitive" communities it is also important to include among those tested hunters and gatherers (who may be absent from a permanent settlement for long periods).   Furthermore, activities must be observed not only at midday, but also during the early morning (when much of the heavy, unstructured work of a tropical community is carried out (Ladell, 1964; Baker, 1966) (Tables 8, 9).

Annual cycles of activity may be imposed not only by climate, but also by months of fasting, particularly in Islamic countries.   In some of the sub-Saharan territories, even multi-year cycles of activity may arise from fluctuations of annual rainfall.

Questionnaires

Activity diaries and questionnaires are relatively simple methods of sampling activity in "western" populations, but become much more difficult to use when the observer lacks familiarity with the language and the culture of a population.

In the "white" population of Canada, a single, well-worded questionnaire accounted for 30% of the difference in aerobic power between those men who are rated as very active and those who are sedentary (Bailey et al, 1976).   More complicated questionnaires (Canada Fitness Survey, 1973) could theoretically examine the energy expenditures associated with various individual activities. However, in practice, many community activities are not to be found in standard catalogues of MET values.   Even where a given task is listed (Kino-Quebec, 1979), difficulties often arise because the individual chooses to pursue the activity at a non-standard rate.

Dietary surveys

A second approach is to estimate the energy content of the food consumed by weighing, interviews, or a 24-hour diary recall (Beaton et al, 1979),

Table 8:   Typical hours of work in "primitive" and intermediate technology societies (based on data collected by Norgan & Ferro-Luzzi, 1978).

Population	Typical hours of work or energy expenditure
Australian aboriginals	Sitting or lying 17 hrs.day^{-1}
Kalahari bushmen	Hunting or foraging 2-3 days.week^{-1}
Dobe bushmen (S.W. Africa) Hazda (Tansania)	2-3 hrs.day^{-1} seeking food
Philippines	1.5-3.5 hrs.day^{-1} cultivating rice
New Guinea	1-2 hrs.day^{-1} cultivating crops
Cameroun peasant farmers	2.5-4.5 hrs.day^{-1}, rising to 10 hrs.day^{-1} at key seasons
Cash crop employees Central America Columbia (sugar cane)	15.4 MJ.day^{-1} 14.3 MJ.day^{-1}

Table 9   Daily energy expenditure of selected tropical populations. Based in part on data accumulated by Shephard (1978), Norgan & Ferro-Luzzi, 1978)

Population	Daily energy expenditure (MJ)
Columbian sugar cane cutters	14.3
Guatemalan farmers	15.4
Indian cotton mill workers	12.8
Jamaican hill farmers	13.6
New Guinea - coastal villagers	11.0
highlanders	10.7

recognising that in the short term energy intake may not match the output in daily activities (Edholm, 1977).

This method has yielded improbably low estimates of daily energy usage in some tropical communities (Table 10). Presumably, the observers concerned had difficulty in collecting complete data, due to language problems and other cultural barriers. In "primitive" societies, portion sizes may be poorly standardised, or wrongly estimated. Parts of the diet (for example, unfamiliar fauna and flora) are often overlooked by an outsider, and estimates of digestibility or energy yield may not apply to local varieties of seemingly familiar foods (Ferro-Luzzi, 1982).

Table 10:   Estimated energy intake of adults living in tropical or sub-tropical
                 regions (based in part on data accumulated by Andersen et al, 1978)

Population	Daily energy intake (kJ)
Baganda peasants, Uganda	8.4
Brazilian villagers	10.5
Cook Island Maoris –	
Men	7.7-9.2
Women	7.4-8.3
East Javan villagers	4.2-6.3
Indian tribes	7.8-13.0
Israeli (Negev bedouin)	5.7-11.1
Kalahari bushmen	8.9
New Guinean villagers –	
Men	9.2-9.6
Women	7.0-7.1
Nigeria (Ibo)	9.8
Philippine villagers	7.0

Since observations cover only a short period, account may need to
be taken of deliberate fasting (Weymes, 1982) or of gorging following starvation.
Inter-individual variability of behaviour is considerable in a "primitive" society,
and it is necessary to follow a large sample of subjects for a long time if reliable
results are to be obtained (Beaton et al, 1979).

Mechanical devices

Pedometers, actometers and accelerometers have occasionally been
used to examine activity patterns in developed societies (Morris, 1973;
McPartland et al, 1975; Saris & Binkhorst, 1977; Veerschuur & Kemper, 1980).
Such devices are best suited to occupations that involve steady walking at a
predictable pace and stride length. Many artefacts would be likely if attempts
were made to record the varied activity patterns of a "primitive" community.

Heart rate recording

Measurements of the exercise heart rate continue to be a popular
method of estimating the intensity of metabolism and/or the stress of effort
(Rodahl, 1977; Rau, 1980). However, the necessary equipment (a tape-
recording monitor or a multi-level heart rate integrator) remains expensive
(Wolff, 1966; Baker et al, 1967; Seliger et al, 1970; Mansourian et al, 1975;
Glagov et al, 1976; Saris et al, 1977). This "high" technology approach is
usually reserved for a select few subjects, and unfortunately they often prove to
be unrepresentative of their community. Furthermore, hot environments cause

problems of data collection, including a tachycardia (which distorts the relationship between metabolism and heart rate) and sweating (which may affect the contact resistance of the ECG electrodes).

Oxygen consumption

The standard approach to the measurement of habitual activity remains the coupling of oxygen consumption measurements (by Kofranyi-Michaelis respirometer, Miser, Oxylog or Medilog recorders; Kofranyi & Michaelis, 1949; McKinnon, 1974; Humphrey & Wolff, 1977; Elay et al, 1976; Andersen et al, 1980) to observer accounts of how subjects spend a "typical" day. Small computers can be programmed to tabulate and integrate the number of minutes subjects spend at each of a series of pre-defined tasks. As with measurements of food intake, there are difficulties in obtaining oxygen-cost data for all tasks, in getting representative values for different seasons and different pursuits within a season, and allowing for inter-individual differences in the speed with which a given activity is performed.

Rating of exertion

A final approach which has found substantial application in developed countries over the past decade has been the rating of perceived exertion (Borg, 1971). In general, it seems a "robust" measure of the intensity of physical activity, with little difference in response patterns for different ethnic groups. However, it does not yet appear to have been applied in a tropical environment. Moreover, at least one report has suggested that this exertion rating may underestimate stress in the heat (Gamberale & Holmer, 1977).

Effect of community nutrition

An inadequate supply of food may restrict not only immediate activity patterns, but also (by its impact upon the learning experiences of the child) long-term physical activity habits (Heywood & Latham, 1971; Apfelbaum et al, 1971). However, quite severe anaemia seems compatible with normal daily activity in some communities (for example, a Brazilian fishing village where haemoglobin levels averaged 8.8 g.dl^{-1}, Gandra & Bradfield, 1971).

LESSONS FROM THE INTERNATIONAL BIOLOGICAL PROGRAMME

Can the Decade of the Tropics learn any practical lessons from problems that were encountered during the I.B.P. studies?

One immediate cause for reflection is the intense difficulty encountered in persuading individual scientists to confirm to an internationally standardised methodology. The I.B.P. (Weiner & Lourie, 1981), the International Committee for Physical Fitness Research (Larson, 1974), the World Health Organisation (Andersen et al, 1971), the International Labour Organisation (Cotes, 1966), the International Council of Sports Medicine and Physical Education (Mellerowicz & Smodlaka, 1981), the Council of Europe (Klissouras & Tokmadis, 1982), the American College of Sports Medicine (1975), several U.S. medical groups (Erb, 1970; Kattus, 1972) and the Canadian Association of Sports Sciences (MacDougall et al, 1983) have all published "standard" exercise test protocols which differ in minor details from one another.   Many individual investigators have further modified these various proposals, and some have ignored them completely.   The reasons for the continuing diversity of practice include ignorance, local pride, and in some cases a shortage of necessary equipment. However, if findings during the "Decade of the Tropics" are to be generalised, data must be comparable from one laboratory to another.   Biological scientists must thus consider how the concordance of methodology typical of the physical sciences can best be achieved.

A second important lesson is that most human adaptability projects involve quite large inter-disciplinary teams, with budgets of corresponding dimensions.   Much of the money that was allocated for I.B.P. research apparently "disappeared" in personal side-projects and incomplete investigations. Often, the vital inter-disciplinary synthesis was not possible because key pieces of information had not been collected.   More thought will thus be needed on methods of selecting project leaders - investigators who can inspire colleagues to work towards a common goal, completing assigned components of a coordinated research programme effectively and on schedule.   Possibly, future teams should include at least one professional research manager.

Finally, greater attention should be paid to insights from the social sciences. We must acknowledge that human kind is adapting to a habitat which itself is undergoing rapid cultural change.   This process influences not only the data that we collect, but also the questions that we ask, and the interpretations that we place upon our findings.

REFERENCES

American College of Sports Medicine (1975).   Guidelines for Graded Exercise Testing and Exercise Prescription, 2nd edn. Philadelphia: Lea & Febiger.
Andersen, K.L., Masironi, R., Rutenfranz, J. & Seliger, V. (1978).   Habitual physical activity and health.   Copenhagen: World Health Organisation Regional Publication **6**, pp. 1-88.

Anderson, K.L., Shephard, R.J., Denolin, H., Varnauskas, E. & Masironi, R. (1971). Fundamentals of Exercise Testing. Geneva, Switzerland: World Health Organisation.

Andreano, R.L. (1976). The recent history of parasitic disease in China: the case of schistosomiasis, some public health and economic aspects. International Journal of Health Services, **6**, 53-68.

Apfelbaum, M., Bostsarron, J. & Lacatus, D. (1971). Effect of caloric restriction and excessive caloric intake on energy expenditure. American Journal of Clinical Nutrition, **24**, 1405-1409.

Arnett, B. & Rikli, R. (1981). Effects of method of subject selection (volunteer vs random) and treatment variable on motor performance. Research Quarterly, **52**, 433-440.

Austin, D.M., Ghesquiere, J. & Azama, M. (1979). Work capacity and body morphology of Bantu and Pygmoid groups of Western Zaire. Human Biology, **50**, 79-89.

Baecke, J.A.H. (1982). Determinants of Body Fatness in Young Adults living in a Dutch Community. Ph.D. thesis, Wageningen, Holland.

Bailey, D.A., Shephard, R.J. & Mirwald, R. (1976). Validation of a self-administered home test of cardiorespiratory fitness. Canadian Journal of Applied Sports Science, **1**, 67-78.

Baker, J.A.S., Humphrey, J.E. & Wolff, H.S. (1967). Socially acceptable monitoring instrument (SAMI). Journal of Physiology (London), **188**, 4-5P.

Baker, P.T. (1966). Ecological and physiological adaptation in indigenous South Americans, In: P.T. Baker & J.S. Weiner (eds.), The Biology of Human Adaptability, pp. 275-303. Oxford: Clarendon Press.

Bar-Or, O. (1983). Paediatric sports medicine for the practitioner. New York: Springer Verlag.

Baum, E., Bruch, K. & Schwennicke, P. (1976). Adaptive modifications in the thermoregulator system of long-distance runners. Journal of Applied Physiology, **40**, 404-410.

Beaton, G.H., Milner, J., Corey, P., McGuire, V., Cousins, M., Stewart, E., de Ramos, M., Hewitt, D., Grambsch, P.V., Kassim, N. & Little, J.A. (1979). Sources of variance in 24-hour dietary recall data: implications for nutrition study design and interpretation. American Journal of Clinical Nutrition, **32**, 2546-2559.

Bliss, C. & Stern, N. (1982). Consumption-work decisions by producers and labourers, In: G.A. Harrison (ed.), Energy and Effort, pp. 159-163. London: Taylor & Francis.

Bogin, B. & MacVean, R.B. (1981). Nutritional and biological determinants of body fat patterning in urban Guatemalan children. Human Biology, **53**, 259-268.

Borg, G. (1971). The perception of physical performance, In: R.J. Shephard (ed.), Frontiers of Fitness, pp. 280-294. Springfield, Illinois: Thomas.

Bouchard, C. (1978). Genetics, growth and physical activity, In: F. Landry & W.A.R. Orban (eds.), Physical Activity and Human Well-being. Miami: Symposia Specialists.

Bouchard, C., Thibault, M.-C. & Jobin, J. (1981). Advances in selected areas of human work physiology. Yearbook of Physical Anthropology, **24**, 1-36.

Brandt, W. (1980).   North-South - A Programme for Survival.   London: Pan Books.

Breymeyer, A.I. & Van Dyne, G.M. (1979).   Grasslands, Systems Analysis and Man. London: Cambridge University Press.

Caiozzo, V.J. & Kyle, C.R. (1980).   The effect of external loading upon power output in stair climbing.   European Journal of Applied Physiology, **44**, 217-222.

Canada Fitness Survey (1983).   Ottawa: Directorate of Fitness and Amateur Sport.

Caplan, A. (1964).   Critical analysis of collapse in underground workers on the Kolar Gold Field.   Transactions of the Institute of Mining Metalurgists, **53**, 95-179. Cited by Ladell (1964).

Cavalli-Sforza, L.L. & Bodmer, W.C. (1971).   The Genetics of Human Populations. San Francisco: Freeman.

Clark, C. & Haswell, M.R. (1964).   The Economics of Subsistence Agriculture. London: MacMillan.

Collins, K.J. (1982).   Energy expenditure, productivity and endemic disease, In: G.A. Harrison (ed.), Energy and Effort, pp. 65-84.   London: Taylor & Francis.

Conconi, F., Ferrai, M., Ziglie, P.G., Droghetti, P. & Codeca, L. (1982). Determination of the anaerobic threshold by a non-invasive field test in runners. Journal of Applied Physiology, **52**, 869-873.

Cotes, J.E. (1966).   Occupational Health and Safety Series.   Report 6.   Geneva ILO.

Cotes, J.E. (1974).   Genetic factors affecting the lung.   Genetic component of lung function. Bulletin de Physio-Pathologie et Respiration, Nancy, **10**, 109-117.

Cotes, J.E. & Davies, C.T.M. (1969).   Factors underlying the capacity for exercise: a study in physiological anthropometry.   Proceedings of the Royal Society of Medicine, **62**, 620-624.

Cotes, J.E., Reed, J.W. & Mortimore, I.L. (1982).   Determinants of capacity for physical work, In: G.A. Harrison (ed.), Energy and Effort, pp. 39-64. London: Taylor & Francis.

Cureton, K.J. (1983).   Error in estimating percentage body fat.   Canadian Journal of Applied Sports Science, **8**, 53.

Davies, C.T.M. (1973).   The relationship of maximum aerobic power output to productivity and absenteeism of East African sugar cane workers. British Journal of Industrial Medicine, **30**, 146-154.

Davies, C.T.M. (1974).   The relationship of leg volume (muscle plus bone) to maximal aerobic power output on a bicycle ergometer; the effects of anaemia, malnutrition and physical activity. Annals of Human Biology, **1**, 47-55.

Davies, C.T.M., Banres, C., Fox, R.H., Ojikutu, R.O. & Samueloff, A.S. (1972). Ethnic differences in physical working capacity.   Journal of Applied Physiology, **33**, 726-732.

Davies, C.T.M., Barnes, C. & Sargeant, A.J. (1971).   Body temperatures in exercise.   Effect of acclimatization to heat and habituation to work. Int.Z.Angew.Physiol. **30**, 10-19.

Davies, C.T.M., Mbelwa, D. & Dore, C. (1974). Physical growth and development of urban and rural East African children, aged 7-16 years. Ann.Hum.Biol. **1**, 257-268.

Davies, C.T.M. & Van Haaren, J.P.M. (1973). Maximum aerobic power and body composition in healthy East African older male and female subjects. American Journal of Physical Anthropology, **39**, 395-401.

Davies, C.T.M. & Van Haaren, J.P.M. (1974). Heart volume in relation to maximal aerobic power output in young anaemic and normal African subjects. Proceedings of the 6th International Symposium on Pediatric Work Physiology, Seč, Czechoslovakia.

Dedoyard, E. & Ghesquiere, J. (1980). Evaluation of aerobic power and physical working capacity of female and male Zairians. In: M. Ostyn, G. Beunen & J. Simons (eds.), Kinanthropometry, pp. 129-141. Baltimore: University Park Press.

DiPrampero, P.E. (1971). Anaerobic capacity and power, In: R.J. Shephard (ed.), Frontiers of Fitness. Springfield, Illinois: Thomas.

Duncan, M.T. & Scammell, C.A. (1977). Physical work capacity and pulmonary function of Malaysian adolescent females. Human Biology, **49**, 31-40.

Edholm, O.G. (1977). Energy balance in man. Journal of Human Nutrition, **31**, 413-431.

Ekblom, B. & Gjessing, E. (Maximal oxygen uptake of the Easter Island population. Journal of Applied Physiology, **25**, 124-129.

Eley, C., Goldsmith, R., Layman, D. & Wright, B.M. (1976). A miniature indicating and sampling respirometer (Miser). Journal of Physiology, **256**, 59-60.

Erb, B.D. (1970). Physician's handbook for evaluation of cardiovascular and physical fitness. Nashville: Tennessee Heart Association, Physical Exercise Committee.

Eveleth, P.B. (1979). Population differences in growth: environmental and genetic factors, In: F. Falkner & J.M. Tanner (eds.), Human Growth, vol. 3: Neurobiology and Nutrition, p. 381. London: Ballière Tindall.

Faria, I., Dix, C. & Frazier, C. (1978). Effect of body position during cycling on heart rate, pulmonary ventilation, oxygen uptake and work output. Journal of Sports Medicine and Physical Fitness, **18**, 49-56.

Ferro-Luzzi, A. (1982). Meaning and constraints of energy intake studies in free-living populations. In: G.A. Harrison (ed.), Energy and Effort, pp. 115-137. London: Taylor & Francis.

Fitzgerald, P.I., Knowlton, R.G., Sedlock, D.A., Tahamont, M.V. & Schneider, D.A. (1980). A comparison of maximal aerobic power predicted from the Canadian Home Fitness Test and a direct treadmill test. Medical Science and Sports, **12**, 88.

Forbes, G.B. (1974). Stature and lean body mass. American Journal of Clinical Nutrition, **27**, 595-602.

Gamberale, F. & Holmer, I. (1977). Heart rate and perceived exertion in simulated work with high heat stress, In: G. Borg (ed.), Physical Work and Effort, pp. 323-332. Oxford: Pergamon Press.

Gandra, Y.R. & Bradfield, R.B. (1971). Energy expenditure and oxygen handling efficiency of anemic schoolchildren. American Journal of Clinical Nutrition, **24**, 1451-1456.

Gardner, G.W., Edgerton, V.R., Senewiratne, B., Barnard, R.J. & Ohira, Y. (1977). Physical work capacity and metabolic stress in subjects with iron deficiency anemia. American Journal of Clinical Nutrition, **30**, 910-917.

Garrison, R.J., Feinleib, M., Castelli, W.P. & McNamara, P.M. (1983). Cigarette smoking as a confounder of the relationship between relative weight and long-term mortality. The Framingham Heart Study. J.A.M.A.. **249**, 2199-2203.

Genverio, S.E. & Dolgener, F.A. (1980). The relationship of isokinetic torque at two speeds to the vertical jump. Research Quarterly, **51**, 593-598.

Ghesquiere, J.L.A. (1971). Physical development and working capacity of Congolese. In: R. Vorster (ed.), Human Biology and Environmental Change. Malawi, I.B.P.

Ghesquiere, J. & Eekels, R. (1984). Fitness of children in Kinshasa. In: J. Ilmarinen & I. Välimäki (eds.), Children and Sport. Berlin: Springer Verlag.

Gitin, E.L., Olerud, J.E. & Carroll, H.W. (1974). Maximal oxygen uptake based on lean body mass: a meaningful measure of physical fitness. Journal of Applied Physiology, **36**, 757-760.

Glagov, S., Rowley, D.A., Cramer, D.B. & Page, R.G. (1970). Heart rates during 24 hours of usual activity for 100 normal men. Journal of Applied Physiology, **29**, 799-805.

Glick, S. & Schvartz, E. (1974). Physical working capacity of young men of different ethnic groups in Israel. Journal of Applied Physiology, **37**, 22-26.

Gold, A.J., Zornitzer, A. & Samueloff, S. (1969). Influence of season and heat on energy expenditure during rest and exercise. Journal of Applied Physiology, **27**, 9-12.

Goldrick, R.B., Sinnett, P.F. & Whyte, H.M. (1970). An assessment of coronary heart disease and coronary risk factors in a New Guinea highland population. In: R.J. Jones (ed.), Atherosclerosis, pp. 366-373. Berlin: Springer Verlag.

Greksa, L.P. & Baker, P.T. (1982). Aerobic capacity of modernizing Samoan man. Human Biology, **54**, 777-799.

Harris, D.R. (1982). Resource distribution and foraging effort in hunter-gatherer subsistence. In: G.A. Harrison (ed.), Energy and Effort, pp. 187-208. London: Taylor & Francis.

Harrison, G.A. & Walsh, R.J. (1974). A discussion on human adaptability in a tropical ecosystem. An I.B.P. human biological investigation of two New Guinea communities. Philosophical Transactions, London (B), **268**, 221-400.

Hawkes, K. & O'Connell, J.F. (1981). Affluent hunters? some comments in the lights of the Alyawara case. American Anthropologist, **83**, 622-626.

Haywood, P.F. & Latham, M.C. (1971). Use of the SAMI heart-rate integrator in children. American Journal of Clinical Nutrition, **24**, 1446-1450.

Hiernaux, J. (1966). Peoples of Africa from 22°N to the Equator, In: P.T. Baker & J.S. Weiner (eds.), The Biology of Human Adaptability. Oxford: Clarendon Press.

Huizinga, J. & Hooijen-Bosma, E.G. (1980). Hemoglobin concentration and physical fitness, In: M. Ostyn, G. Beunen & J. Simons (eds.), Kinanthropometry II. Baltimore: University Park Press.

Humphrey, S.J.E. & Wolff, H.S. (1977). The Oxylog. Journal of Physiology, **267**, 12P.

Johnston, F.E., Bogin, B., MacVean, R.B. & Newman, B.C. (1984). A comparison of international standards versus local reference data for the triceps and subscapular skinfolds of Guatamalan children and youth. Human Biology, **56**, 157-171.

Jones, N. & Kane, M. (1979). Interlaboratory standardization of data. Medicine and Science in Sports, **11**, 368-372.

Jones, P.R.M. & Corlett, J.T. (1980). Some factors affecting the calculation of human body density: bone demineralization. In: M. Ostyn, G. Beunen & J. Simons (eds.), Kinathropometry II. Baltimore: University Park Press.

Jones, P.R.M. & Pearson, J. (1970). Anthropometric determination of leg fat and muscle plus bone volumes in young male and female adults. Journal of Physiology, **204**, 63-64P.

Kattus, A.A. (1972). Exercise testing and training of apparently healthy individuals: a handbook for physicians. New York: American Heart Association.

Kay, C. & Shephard, R.J. (1969). On muscle strength and the threshold of anaerobic work. Int.Z.Angew.Physiol. **27**, 311-328.

Kelly, J.M., Serfass, R.C. & Stull, G.A. (1980). Elucidation of maximal oxygen uptake from standing bicycle ergometry. Research Quarterly, **51**, 315-322.

King, D.S., Brodowicz, G.R. & Ribisl, P.M. (1982). The effect of toe clip use on maximal oxygen uptake during bicycle ergometry in competitive cyclists and trained non-cyclists. Medicine and Science in Sports, **14**, 147.

Kino-Québec (1979). Groupe d'étude de Kino-Québec sur le système de quantification de la dépense energétique (GSQ). Annexe to Final Report.

Klissouras, V. & Tokmakidis, S. (1982). Evaluation of physical fitness of schoolchildren. The Eurofit test, In: H. Howald (ed.), Youth and Sports. Magglingen: Swiss School for Physical Education and Sports.

Kofranyi, E. & Michaelis, H.F. (1949). Ein tragbarer Apparat zur Bestimmung des Gastoffwechsels. Int.Z.Angew.Physiol. **11**, 148-150.

Kofsky, P., Davis, G. & Shephard, R.J. (1983). Muscle strength and aerobic power of the lower-limb disabled. Ann.del Istituto Sup.di Ed.Fis. **2**, 201-208.

Ladell, W.S.S. (1951). Assessment of group acclimatization to heat and humidity. Journal of Physiology, **115**, 296-312.

Ladell, W.S.S. (1964). Terrestrial animals in humid heat: man, In: D.B. Dill (ed.), Handbook of Physiology, Section 4. Washington D.C.: American Physiological Society.

Ladell, W.S.S. & Shephard, R.J. (1961). Aldosterone inhibition and acclimatization to heat. Journal of Physiology (London), **160**, 19-20P.

Larson, L.A. (ed.) (1974). Fitness, Health and Work Capacity. International Standards for Assessment. New York: MacMillan.

LaVoie, N., Dallaire, J., Brayne, S. & Barrett, D. (1984). Anaerobic testing using the Wingate and Evans-Quinney protocols with and without toe stirrups. Canadian Journal of Applied Sports Science, **9**, 1-5.

Lee, R.B. (1969). !Kung Bushmen subsistence: an input-output analysis, In: A. P. Vayda (ed.), Environment and Cultural Behaviour. New York: Natural History Press.

Lloyd, B.B. (1966).   Presidential address, Section 1 (Physiology and Biochemistry), British Association,   In: Advancement of Sciences, pp. 515-530.

Lohman, T.G. (1981).   Skinfolds and body density and their relation to body fatness: a review. Human Biology, **53**, 181-226.

MacDougall, J.D., Reddan, W.G., Layton, C.R. & Dempsey, J.A. (1974).   Effects of metabolic hyperthermia on performance during heavy, prolonged exercise. Journal of Applied Physiology, **36**, 538-544.

MacDougall, J.D., Wenger, H.A. & Green, H.J. (1983).   Physiological testing of the elite athlete. Ottawa: Canadian Association of Sport Sciences.

MacPherson, R.K. (1966).   Physiological adaptation, fitness and nutrition in the peoples of the Australian and New Guinea regions. In: P. T. Baker & J. S. Weiner (eds.),   The Biology of Human Adaptability.   Oxford: Clarendon Press.

Malhotra, M.S. (1966).   People of India, including primitive tribes - a survey on physiological adaptation, physical fitness and nutrition. In: P.T. Baker & J.S. Weiner (eds.),   The Biology of Human Adaptability.   Oxford: Clarendon Press.

Malina, R.M. (1984).   Motor development and performance of children and youth in undernourished populations.   16th Scientific Meeting of International Committee for Physical Fitness Research, Eugene, Oregon.

Mann, G.V., Munoz, J.A. & Scrimshaw, N.S. (1955).   The serum lipoprotein and cholesterol concentrations of central and north Americans with different dietary habits. American Journal of Medicine, **19**, 25-32.

Margaria, R. (1966).   An outline for setting significant tests of muscular performance,   In: H. Yoshimura & J.S. Weiner (eds.),   Human Adaptability and its Methodology. Tokyo: Society for the Promotion of Sciences.

Mathur, D.N. (1982).   Body composition of top-ranking Nigerian basketball players.   South African Journal of Sport, Physical Education and Recreation, **5**, 15-20.

Mansourian, P., Masironi, R., Nicoud, J.D. & Steffen, P. (1975).   Recording the cardiac interbeat interval distribution.   Journal of Applied Physiology, **38**, 542-545.

McKinnon, J.B. (1974).   Miniature 4-channel cassette recorder for physiological and other variables. In: P. A. Neukomm (ed.), Biotelemetry II. Basel: Karger.

McPartland, R., Kupfer, D.J., Foster, F., Reisler, K. & Matthews, G. (1974).   Objective measurement of human motor activity: a preliminary normative study. Biotelemetry **2**, 317-323.

Mellerowicz, H. & Smodlaka, V.N. (1981).   Ergometry.   Basics of Medical Exertion. Baltimore: Urban & Schwarzenburg.

Metropolitan Life Assurance Company (1983).   Nutrition and Athletic Performance. New York: Metropolitan Life Assurance Company.

Miller, G.J., Cotes, J.E., Hall, A.M., Salvosa, C.B. & Ashworth, A. (1972).   Lung function and exercise performance of healthy Caribbean men and women of African ethnic origin. Quart.J.Exp.Physiol. **57**, 325-341.

Morris, J.R.W. (1973).   Accelerometry - a technique for the measurement of human body movements. Journal of Biomechanics, **6**, 729-736.

Montoye, H.J. & Gayle, R. (1978). Familial relationships in maximal oxygen uptake. Human Biology, **50**, 241-250.

Norgan, N.G. & Ferro-Luzzi, A. (1978). Nutrition, physical activity and fitness in contrasting environments. In: J. Parízková & V.A. Rogozkin (eds.), Nutrition, Physical Fitness and Health, pp. 167-193. Baltimore: University Park Press.

Norris, N.G. (1982). Human energy stores, In: G.A. Harrison (ed.), Energy and Effort, pp. 139-158. London: Taylor & Francis.

Nottrodt, J.W. & Celentano, E.J. (1984). Use of validity measures in the selection of physical screening tests, In: D.A. Attwood & C. McCann (eds.), Proceedings of the 1984 International Conference on Occupational Ergonomics, vol. 1, pp. 433-437. Toronto: Human Factors Association of Canada.

Parízková, J., Merhautova, J. & Prokopec, M. (1972). Comparaison entre la croissance des jeunes Tunisiens et celle des Jeunes Tchéques âges de 11 et 12 ans. Biometr.Hum.(Paris) **7**, 1-10.

Parkhouse, W.S., McKenzie, D.S., Rhodes, E.C., Dunwoody, D. & Wiley, P. (1982). Cardiac frequency and anaerobic threshold: implications for prescriptive exercise programs. European Journal of Applied Physiology, **50**, 117-123.

Patrick, J.M. (1976). Respiratory responses to CO_2 rebreathing in Nigerian men. Quarterly Journal of Experimental Physiology, **61**, 85-93.

Patrick, J.M. & Cotes, J.E. (1971). Cardiac determinants of aerobic capacity in New Guineans. In: V. Seliger (ed.), Physical Fitness. Prague, Czechoslovakia: Charles University Press.

Patrick, J.M. & Cotes, J.E. (1978). Cardiac output during submaximal exercise in New Guineans: the relation with body size and habitat. Quarterly Journal of Experimental Physiology, **63**, 277-290.

Petrofsky, J.S. & Phillips, C.A. (1981). The influence of body fat on isometric exercise performance. Ergonomics **24**, 215-222.

Pirnay, F., Deroanne, R. & Petit, J.M. (1970). Maximal oxygen consumption in a hot environment. Journal of Applied Physiology, **28**, 642-645.

Powers, S.K., Dodd, S., Deason, R., Byrd, R. & McKnight, T. (1983). Ventilatory threshold, running economy and distance running performance of trained athletes. Research Quarterly, **54**, 179-182.

Prior, I.A.M. & Evans, J.G. (1970). Current developments in the Pacific, In: R.J. Jones (ed.), Atherosclerosis, pp. 335-342. Berlin: Springer Verlag.

Purvis, J.W. & Cureton, K.J. (1981). Ratings of perceived exertion at the anaerobic threshold. Ergonomics, **24**, 295-300.

Quinney, H.A., Cottle, W.H., McDougall, A. & Baleshta, J. (1983). The effect of temperature variation on cardiovascular response to the STF step test. Canadian Journal of Applied Sport Science, **8**, 206.

Rau, G. (1980). Electrophysiological measurement techniques. In: Manned Systems Design: new methods and equipment, pp. 181-198. NATO Defence Research Section I. New York: Plenum.

Ready, A.E. & Quinney, H.A. (1982). Alterations in anaerobic threshold as the result of endurance training and detraining. Medicine and Science in Sports, **14**, 292-296.

Richards, P. (1982).   Quality and quantity in agricultural work - Sierra Leone rice farming systems,   In: G.A. Harrison (ed.),   Energy and Effort, pp. 209-228. London: Taylor & Francis.

Rivers, J.P.W. & Payne, P.R. (1982). The comparison of energy supply and need: a critique of energy requirements.   In: G.A. Harrison (ed.), Energy and Effort, pp. 85-105. London: Taylor & Francis.

Robertshaw, S.A., Reed, J.W., Mortimore, I.L., Cotes, J.E., Afacan, A.S. & Grogan, J.B. (1984). Submaximal alternatives to the Harvard Pack Index as guides to maximal oxygen uptake (physical fitness). Ergonomics **27**, 177-185.

Rodahl, K. (1977).   On the assessment of physical work stress,   In: G. Borg (ed.), Physical Work and Effort, pp. 199-216. Oxford: Pergamon Press.

Ross, W.D., Drinkwater, D.T., Bailey, D.A., Marshall, G.W. & Leahy, R.M. (1978). Kinanthropometry - traditions and new perspectives,   In: M. Ostyn, G. Beunen & J. Simons (eds.),   Kinanthropometry II, pp. 3-32. Baltimore: University Park Press.

Rowell, L.B. (1974).   Human cardiovascular adjustments to exercise and thermal stress. Physiological Reviews, **54**, 75-159.

Rowell, L.B., Brengelmann, G.L., Blackmon, J.R., Twiss, R.D. & Kusumi, F. (1968).   Splanchnic blood flow and metabolism in heat-stressed men. Journal of Applied Physiology, **24**, 475-484.

Ruskin, H. (1978).   Physical performance of school children in Israel.   In: R.J. Shephard (ed.),   Physical Fitness Assessment: principles, practice and applications. Springfield, Illinois: Thomas.

Saltin, B., Henriksson, J., Nygaard, E. & Andersen, P. (1977).   Fiber types and metabolism potentials of skeletal muscules in sedentary man and endurance runners.   Annals of the New York Academy of Science, **301**, 3-29.

Saris, W.H.M. & Binkhorst, R.A. (1977). The use of pedometer and actometer in studying daily physical activity. European Journal of Applied Physiology, **37**, 219-228.

Saris, W.H.M., Snell, P. & Binkhorst, R.A. (1977).   A portable heart rate distribution recorder for studying daily physical activity.   European Journal of Applied Physiology, **37**, 17-25.

Satwanti, K., Bharadwaj, H. & Singh, I.P. (1977). Relationship of body density to body measurement in young Punjabi women: applicability of body composition prediction equations developed for women of European descent. Human Biology, **49**, 203-213.

Satyanarayana, K., Naidu, A.N., Chatterjee, B. & Rao, B.S.N. (1978). Nutritional status, body physique and work output,   In: J. Pařízková & J. Rogozkin (eds.),   Nutrition, Physical Fitness and Health, pp. 215-220. Baltimore: University Park Press.

Seliger, V., Hrdlicka, J., Kokes, A. & Zelenka, K. (1970).   Zarizeni pro dlouhodobe sledovani pohybove aktivity cloveka.   The device for long-lasting investigations of moving activity in man.   Cs. Fysiol. **19**, 269-273.

Senay, L.C. & Kok, R. (1977).   Effects of training and heat acclimatization on blood plasma controls. Journal of Applied Physiology, **43**, 591-599.

Shephard, R.J. (1977).   Endurance Fitness, 2nd edn.   Toronto, Ontario: University of Toronto Press.

Shephard, R.J. (1978). Human Physiological Work Capacity. London: Cambridge University Press.

Shephard, R.J. (1982). Physiology and Biochemistry of Exercise. New York: Praeger.

Shephard, R.J. (1984). Tests of maximum oxygen intake: a critical review. Sports Medicine 1, 99-124.

Shephard, R.J., Lavallée, H., LaBarre, R., Jéquier, J.-C., Volle, M. & Rajic, M. (1980). The basis of data standaradization in prepubescent children, In: M. Ostyn, G. Beunen & J. Simons (eds.), Kinanthropometry II, pp. 360-370. Baltimore: University Park Press.

Sinnett, P.F. & Whyte, H.M. (1973). Epidemiological studies in a highland population of New Guinea. Environment, culture and health status. Human Ecology 1, 245-277.

Sodhi, H.S. & Sidhu, L.S. (1984). Physique and selection of sportsmen. Patialia: Punjab Publishing House.

Spurr, G.B. (1984). Physical activity, nutritional status, and physical work capacity in relation to agricultural productivity. In: E. Pollitt and P. Amante (eds.), Energy Intake and Activity, pp. 207-261. New York: Liss.

Strydom, N.B. (1978). Environmental variables affecting testing, In: R.J. Shephard & H. Lavallée (eds.), Principles, Practice and Applications. Springfield, Illinois: Thomas.

Thinkaren, T., Chan, O.L., Duncan, M.T. & Klissouras, V. (1975). Absence of the influence of ethnic origin on the maximal aerobic power of Malaysians, In: H. Toyne (ed.), World Congress of Sports Medicine. Melbourne: Australian Sports Federatioin.

Thompson, J.M. (1981). Prediction of anaerobic capacity: a performance test employing an optimal exercise stress. Canadian Journal of Applied Sport Science, 6, 16-20.

Thys, H., Dreezen, E. & Vanderstaffen, A. (1979). Influence de la modalité d'exercice sur la valeur de la VO_2 max. Archives Internationales de Physiologie et de Biochimie, 87, 565-573.

U.S. Sports Academy (1983). International Physical Fitness Test. Bahrein: General Organization of Youth and Sport.

Van Ee, J.H. (1985). Work performance of West African children, In: W. Saris, H. Kemper & R. Binkhorst (eds.), Pediatric Work Physiology. Champaign, Illinois: Human Kinetics Publishers.

Veerschuur, R. & Kemper, H.C.G. (1980). Adjustment of pedometers to make them more valid in assessing running. International Journal of Sports Medicine, 1, 87-89.

Viteri, F.E. (1971). Considerations on the effect of nutrition on the body composition and physical working capacity of young Guatemalan adults, In: N.S. Scrimshaw & A.M. Altschul (eds.), Amino Acid Fortification of Protein Foods, pp. 350-375. Cambridge: MIT Press.

Von Döbeln, W. (1966). Kroppstorlek, Energieomsättning och Kondition. In: G. Luthman, U. Åberg & N. Lundgren (eds.), Handbok i Ergonomi. Stockholm: Almqvist & Wiksell, 1966.

Vorster, D.J.M. (1977). Adaptation to urbanization in South Africa, In: G.A. Harrison (ed.), Population Structure and Human Variation, pp. 313-332. London: Cambridge University Press.

Weber, G., Kartodihardjo, W. & Klissouras, V. (1976).   Growth and physical training with reference to heredity.   Journal of Applied Physiology,   **40**, 211-215.

Weiner, J.S. (1950).   Observations on the working ability of Bantu mine workers with reference to acclimatization to hot, humid conditions.   British Journal of Industrial Medicine,   **7**, 17-26.

Weiner, J.S. & Lourie, J.A. (1981).   Practical Human Biology.   London: Academic Press.

Weymes, H. (1982).   Determinants of nutritional need,   In: G.A. Harrison (ed.), Energy and Effort, pp. 107-114.   London: Taylor & Francis.

Williams, C.G., Bredell, G.A.G., Wyndham, C.H., Strydom, N.B., Morrison, J.F., Peter, J., Fleming, P.W. & Ward, J.S. (1962).   Circulatory and metabolic reactions to work in heat.   Journal of Applied Physiology,   **17**, 625-638.

Wolff, H. (1966).   Physiological measurement on human subjects in the field, with special reference to a new approach to data storage.   In: H. Yoshimura & J.S. Weiner (eds.),   Human Adaptability and Its Methodology.   Tokyo: Japanese Society for Promotion of Sciences.

Worthington, E.B. (1978).   The Evolution of I.B.P.   London: Cambridge University Press.

Wyndham, C.H. (1966).   Southern African ethnic adaptation to temperataure and exercise.   In: P.T. Baker & J.S. Weiner (eds.),   The Biology of Human Adaptability, pp. 201-244.   Oxford: Clarendon Press.

Wyndham, C.H. (1967).   Commentary,   In: R.J. Shephard (ed.),   Proceedings of International Symposium on Physical Activity and Cardiovascular Health. Canadian Medical Association Journal,   **96**, 835.

Wyndham, C.H. (1974).   1973 Yant Memorial Lecture: research in the human sciences in the gold mining industry.   American Industrial Hygiene Journal,   **35**, 113-136.

Wyndham, C.H. & Sluis-Cremer, G. (1968).   The capacity for physical work of white miners in South Africa.   II. The rate of oxygen consumption during a step test.   South African Medical Journal,   **42**, 841-844.

VENTILATORY CAPACITY IN TROPICAL POPULATIONS: CONSTITUTIONAL AND ENVIRONMENTAL INFLUENCES

J. M. PATRICK

Department of Physiology & Pharmacology,
University Medical School, Nottingham, U.K.

INTRODUCTION

Physical working capacity is largely determined by the individual's aerobic capacity or maximal oxygen uptake ($\dot{V}O_2$ max). This in its turn depends upon the capacities of a linked series of oxygen transfer functions: diffusion through tissues, circulation of haemoglobin, but first of all pulmonary ventilation. It is the capacity of this primary process of ventilatory function with particular application to tropical populations which is the subject of this paper.

The capacity of the ventilatory system can be measured directly as maximum voluntary ventilation or maximum sustainable ventilation, but it is commonly assessed using single-breath tests. The most convenient measurements are of forced vital capacity (FVC), which reflects the volume of the thorax and the strength of the respiratory muscles, and the forced expired volume in one second (FEV_1), which depends on the FVC and also upon the calibre of the airways.

The first question to ask is: what relevance do these indices of ventilatory capacity have for working capacity? That can be answered in two ways: first by comparing mean values for ventilatory capacity between populations having widely different aerobic capacities. In the preparatory stages of the IBP Human Adaptability programme, Cotes and Davies (1969) measured the aerobic capacities of three European groups: male and female factory workers and some elite athletes. Their $\dot{V}O_2$ max was 2.2, 3.4 and 4.6 l/min respectively. Plotted against this in Figure 1 are four elements of the oxygen transfer chain: vital capacity, transfer factor (formerly diffusing capacity), total Hb, and the reciprocal of cardiac frequency at a sub-maximal work-rate, and in addition two indices of the size of the muscle mass that performs the external work. All these variables are directly related to the

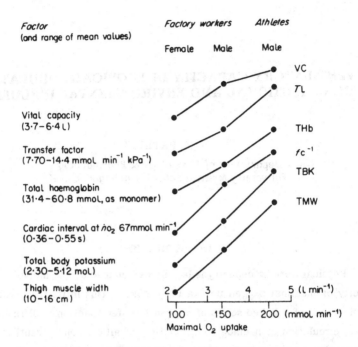

Figure 1:   Four variables relating to the capacity of the oxygen transport
               chain and two variables relating to the size of body muscle, plotted
               against aerobic capacity in three European subject groups. (Data
               of Cotes, reproduced by permission of the author and of Blackwell
               Scientific Publications.)

$\dot{V}O_2$ max, each with a range of values in proportion to the $\dot{V}O_2$ max itself. This
parallel pattern of covariation suggests that no single element of the oxygen
transfer chain constitutes a limiting factor: each makes an individual and
important contribution.   The corollary is that an increment in any one of the
functions might be associated with, if not determine, a corresponding increment
or advantage in working capacity.

       The second way of studying this relation is a statistical one.   If $\dot{V}O_2$ max
and all the other oxygen transport functions are measured within a population,
we can estimate how much of the variance in aerobic capacity is accounted for
by variance in each of these functions.   The following multiple regression
equation:

       $\dot{V}O_2$ max = 0.19 thigh muscle width (cm) - 0.011 cardiac frequency at

               $\dot{V}O_2$ = 1 l/min (bt/min) + 0.160 FVC (l) + 0.138

determined by Cotes and Davies (1969) shows that $\dot{V}O_2$ max depends on thigh
muscle width and standardised cardiac frequency, and a significant independent
contribution of ventilatory capacity, represented by FVC.   For every extra litre

Table 1: Factors contributing to variance in ventilatory capacity

---

A. **Constitutional (genetic)**
    1. Age; growth and decline
    2. Size
    3. Sex
    4. Ethnic group
    5. Other heritable factors

B. **Environmental (adaptive)**
    1. Customary physical activity
    2. Altitude of residence
    3. Atmospheric pollution, including occupational exposure
    4. Smoking

C. **Respiratory Disease**

---

of FVC, 160 ml O_2/min is added to the $\dot{V}O_2$ max. These two complementary approaches, the first examining correlations between populations and the second within, provide suggestive evidence that ventilatory capacity at least *contributes* to the determination of physical working capacity. This is not to say, of course, that ventilatory capacity *limits* working capacity, any more than working capacity itself *limits* the daily productivity of a Sudanese cane-cutter or a Nigerian coal-miner. But just as a worker during his quotidian shift tends to pace himself, on average, at about 40% of his aerobic capacity (Levine et al, 1982); and just as he limits his tidal volume to 50% of his vital capacity when his ventilation is elevated by his exertions (Cotes, 1979), so an individual's ventilatory capacity needs to be taken account of - along with his total body haemoglobin, his socio-economic status, and his health - when assessing his physical working capacity.

Table 1 lists some of the main factors that influence an individual's ventilatory capacity and thus, if the hypothesis above holds true, his working capacity. These factors have largely been explored in European and North American populations but the aim of the present paper is to consider their importance in tropical populations. The apparent simplicity of the classification in Table 1 conceals many instances of interaction: between genetic and environmental influences for example;   and environmental effects on ageing. Roberts (1987) has reviewed the twin and family studies demonstrating heritable influences on ventilatory capacity. Ethnic differences, especially between tropical populations, are of particular concern here. Respiratory disease is too broad to encompass here but three environmental factors have been selected for review: occupational exposure to dust, smoking, and customary physical activity. We will start with the latter because it provides an opportunity to establish some methodological guidelines.

PHYSICAL ACTIVITY

The positive relation between ventilatory capacity and aerobic capacity has already been mentioned. Insofar as the latter can be increased or reduced by physical training or deconditioning (e.g. by bed-rest) it can be regarded as an index of a purely adaptive influence on the organism especially if appropriate adjustments are first made for differences in muscle mass which is an independent determinant of aerobic capacity.

Figure 2 taken from the data of Ghesquiere (1972, 1979) suggests that those subjects who by training have raised their maximal oxygen uptake also have larger ventilatory capacities when stature has been allowed for. These data refer to different samples of Bantu subjects studied in Zaire but the interpretation is not altogether clear. The subjects' ages were not given; and the range of statures from the Twa pygmoids at 1.60 m to the national football team at 1.72 m suggests that there may still be an element of genetic diversity within this national population. Furthermore, no allowance has been made for differences in muscle mass. Nevertheless, Figure 2 provides suggestive evidence of an important link between customary physical activity and ventilatory capacity.

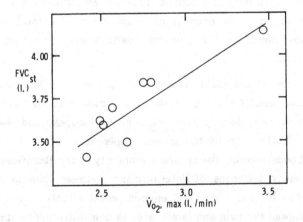

Figure 2:   Forced vital capacity (adjusted to a stature of 1.65 m) plotted against aerobic capacity in 8 groups of male subjects in Zaire. (Data of Ghesquiere.)

Ghesquiere also uses the common practice and expresses $\dot{V}O_2$ max per kilogram of body weight; substitutes for body weight are lean body mass, muscle mass or leg volume, whichever variable has been taken to represent the size of the tissue primarily contributing to the whole-body metabolism. Whatever variable is used, this procedure assumes that there is a direct proportionality between the two, such that the *ratio* is independent of the muscle mass (or volume, etc.). The fallacy of this approach was demonstrated in a classical paper by Tanner (1949). Respiratory physiologists do not adjust for differences in body size between adults by simply dividing vital capacity by stature. Rather the partial regression coefficient on stature is used to adjust vital capacity to what it would have been at some convenient standard value for stature, say, 1.7 m in adult males. Why is $\dot{V}O_2$ max not adjusted to what it would have been at a body weight of, say, 70 kg, using some standard regression coefficient?

An alternative index of the cardiovascular adaptation resulting from physical training or high customary physical activity is the reduction in the heart-rate response to formal sub-maximal exercise, e.g. on an ergometer or in a self-paced walking test (Cotes et al, 1973; Patrick & Cotes, 1978; Ballal et al, 1981). The measurements are highly reproducible, convenient and safe to obtain and they are applicable to larger groups, particularly of older subjects (Patrick et al, 1983; Cotes, 1987, this volume). Figure 3 shows data for FEV_1 (adjusted for age and stature) in four ethnic groups, plotted against such an index of physical training or activity (cardiac frequency at $\dot{V}O_2 = 1.5$ l/min adjusted for fat-free mass). All the measurements were made using the same standard methods. Within each homogeneous group, two subsets differing widely in their physical conditioning are identified. For New Guineans, coastal villagers living a life of ease are contrasted with highlanders eking out a living on steep terrains (Cotes et al, 1974; Norgan et al, 1974). In India lowland Rajput soldiers are compared with Gurkhas from the highlands (Cotes et al, 1975). Afro-Caribbean clerks in Trinidad are compared with farmers in the adjacent hills (Miller et al, 1972), and UK factory workers are compared with a group of amateur racing cyclists (Cotes, 1976). Within each pair the upward slope represents the influence of physical activity on ventilatory capacity as in Figure 2. The residual displacement between the pairs is likely to reflect ethnic differences between these geographically separated populations, as discussed in the next section.

However, in both these examples there remains the possibility that, in some instances at least, individuals who are inately better endowed with higher ventilatory capacities choose, or are chosen, to take on the more active life-styles. This possibility of self-selection seems to have been excluded in the

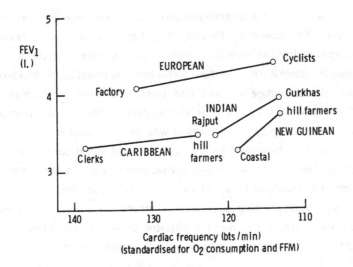

Figure 3: FEV_1 (adjusted to 1.7 m and 25 yrs) plotted against an index of physical condition in men of four ethnic groups each with two subgroups differing in physical condition. (Data of Cotes, Miller, Patrick et al.)

study of a genetically homogeneous group of Chinese children by Jones et al (1977).   Some lived in squatter huts on the hillsides and lived an open-air life. Others were inactive, living in high-rise tenements with lifts but no playgrounds, kept indoors.   The active children, shaded bars in Figure 4, have significantly higher ventilatory capacities, corrected for stature.   This study provides clear evidence that high customary physical activity is a material factor in the growth and development of ventilatory capacity in children, and is entirely consistent with the comparisons cited within and between adult tropical populations above, and with data for Nigerian coal-miners discussed later.

ETHNIC DIFFERENCES IN VENTILATORY CAPACITY

Adults

The data for FEV_1 discussed in the previous section on physical activity and shown in Figure 3 provide a clear example of residual differences in ventilatory capacity that can most probably be attributed to ethnicity. Differences in body size and age were allowed for on the ordinate, and differences in physical training through customary physical activity were expressed along the abscissa which also takes into account the differences in fat-free mass.   Yet at a constant physical conditioning level, indicated by a standardised cardiac

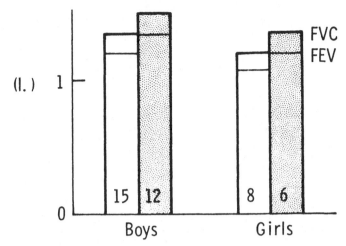

Figure 4: FEV_1 and FVC in boys and girls in Hong Kong. (N shown.) Shaded bars represent active children and unshaded bars inactive children. (Data of Jones et al.)

frequency of 115 bt/min, European subjects had a FEV_1 of 4.4, Indian servicemen 3.8 and New Guineans 3.5 l. (And if we permitted ourselves to extrapolate the Caribbean subjects to a higher physical condition (lower cardiac frequency) than they actually achieved, their FEV_1 would be similar to the New Guineans.) Some differences in ventilatory capacity between different populations can be accounted for by variations in measurement technique: in instrumentation, in the degree of encouragement and practice given to the subject and in the subject's posture. But all those data in Figure 3 were collected using standard IBP methods (Weiner & Lourie, 1969), by observers equipped and trained at a single centre. And there have been numerous other studies in which subjects of different genetic stock have been compared directly in the same environment and differences in ventilatory capacity found: many of these studies are cited in Cotes (1979).

Europeans have the largest values. The rank order of other ethnic groups has not been established, but some information can be obtained from an overview of the studies reported in the world literature. This approach is open to the doubts already expressed about methodology, but the enormous range of subject groups that have been studied provides a compensating advantage. Let us take just African and Indian data as an example, and confine our attention to FVC, the most commonly measured variable. Only if age and stature are quoted by the authors can the data be used, because FVC has to be adjusted to some

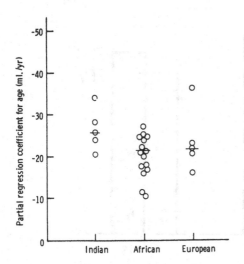

Figure 5:   Rate of age-decline of FVC adjusted for stature in groups of
                Indian, African and European male and female subjects.   Each
                point represents a separate published study.   Median values for
                each ethnic group is shown by a horizontal line.

arbitrary but common values to permit legitimate comparison. Adjustments, as
discussed in the previous section, are made using partial regression coefficients.
But whose coefficients should be used - Europeans'?   Are there differences in
ageing rates between ethnic groups?   Some age coefficients gleaned from the
literature for men are shown in Figure 5;   each symbol represents a separate
study of a sample of between 100 and 5000 subjects. Indians and Africans are
separated and some specimen European studies are included for comparison.

      There is a suggestion that the Indian's rate of decline with age is faster
than the African's, but in view of the wide spread of values no such difference is
firmly established here.   Nor is it confirmed by Miller's direct comparison of
African and Indian migrants in Guyana (1970).   While there have been three
direct comparisons between European and African populations, there are none
between European and Indians and this seems to warrant a new study.
25 ml/year has been used as the standard value for adjustment of means in the
calculations following - not far from the median values for these three groups.

      A similar analysis for stature coefficients, i.e. the relation between FVC
and stature when age is controlled, shows a similar pattern with no striking
differences between groups.   5 l/m has been taken as the standard value.   Note
that this does not mean that a man of 1.7 m has an FVC of 8.5 l (see previous
section).

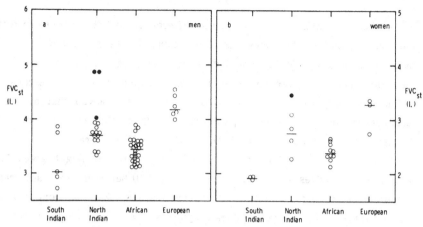

Figure 6:   Forced vital capacity for (a) men (adjusted to 1.65 m and 35 yr) and
(b) women (adjusted to 1.55 m and 35 yr) in groups of North and
South Indian, African and European subjects. Each point
represents a separate published study. Median values are shown.
The filled symbols represent Himalayan populations.

Group mean values of FVC can thus be adjusted to what they would have
been at some convenient arbitrary age and height. For men, 1.65 m and 35
years have been chosen because the commonly used value of 1.7 m is often above
the mean value for height in these tropical populations and 35 years is well into
the age-decline that may not start until 25 years. In Figure 6 each symbol again
represents a single study. Again, the adjusted values for FVC are arranged by
ethnic group, and the medians shown.

The Africans are tightly packed at around 4 l, while the Europeans are
closer to 5 l, a clear ethnic difference. The Indians from north of latitude 22°N
overlap the Africans; the highest points represent the mountain people from the
Himalayas, very active on their steep terrain (Beall, 1987). The South Indians
have some rather low values, and indeed when the studies are weighted by the
number of subjects they are seen to be significantly lower than the North
Indians. This difference cannot be attributed to smoking (see later section).
Figure 6b shows a similar pattern among women (values adjusted to 1.55 m and
35 years) which confirms the difference between Europeans and Africans and
between North and South Indians. It is now important to establish whether or
not such closely related variants of the Indian Caucasoids as these northern and
southern subsets have distinctly different ventilatory capacities, or whether this
apparent "ethnic" difference is attributable to as yet unidentified environmental
factors.

The explanation for these ethnic differences probably resides in differences in body shape, when linear size (i.e. stature) has been allowed for. This is another vindication for the approach known as physiological anthropometry. Shape differences between European and Negro subjects have been described, as in Tanner's study of Olympic athletes (1964).   He showed that among the variables that differ significantly, the trunk of black athletes was shorter than whites at a given stature though the shoulder widths were the same.   This suggests that thoracic volume might be smaller at the same height, giving lower ventilatory capacities in blacks.   Paul et al (1960) had compared radiographic measurements of thoracic dimensions in black and white Zambian copper-miners. Each dimension was smaller in black subjects by 8% to 10%, while stature was only 4% different.   Allowing for this, and multiplying up the linear differences, the thoracic volume is 13% greater in white subjects.   Rossiter & Weil (1974) confirmed that if FVCs in black subjects are first scaled up by 13% they become indistinguishable from those of white subjects.   For this reason, sitting height might be a better index of size than stature when making comparisons between ethnic groups (Hsi et al, 1983), and thoracic dimensions should also be included. Such an analysis, based on the full IBP list of anthropometric measurements, needs to be extended to a comparison of other ethnic groups.   That is, Rossiter and Weil's 13% scaling factor needs to be recalculated for each ethnic group in turn.

One further curious ethnic difference in respiratory function might be noted here.   The size of the vital capacity appears to determine the sensitivity of the ventilatory response to CO_2, a factor of some importance in the pathophysiology of both respiratory failure in adults and possibly also of cot death syndrome in infants.   Figure 7 shows that among five different groups of New Guinean subjects, the CO_2-sensitivity is related to ventilatory capacity in the same way as for Europeans, but with a significantly lower intercept (Patrick & Cotes, 1974).   A group of Nigerian men shared the European response (Patrick, 1976).   This respiratory insensitivity of New Guineans to CO_2, also reported by Beral and Reed (1971), remains unexplained.

Children

If ethnic differences in ventilatory capacity have their origin in body shape differences, then comparison of the development of both skeletal and pulmonary dimensions during childhood might be helpful.   Comparisons of ventilatory capacity have been made between black and white children in North America (Schoenberg et al, 1978; Hsu et al, 1979), between Mexican-American

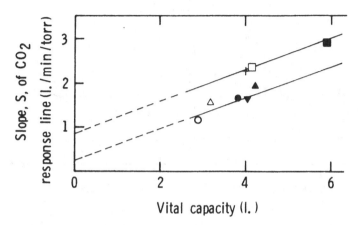

Figure 7:   Ventilatory sensitivity (S) to inhaled CO_2 plotted against vital capacity in eight subject groups. ■ , ☐ European men and women; ●, O Lowland New Guineans; ▲ , △ Highland New Guineans; ▼ Migrant Highland New Guineans at sea level; + Nigerian men. (Data of Patrick & Cotes.)

and white children in North America (Hsi et al, 1983), and between European children and migrants of African and Indian stock in the Caribbean (Miller et al, 1977). These studies demonstrate that the differences in ventilatory capacity between groups of children of different ethnic groups, growing up in the same environment and with apparently similar patterns of activity, are already apparent by the time the children are able to blow into a spirometer, and the differences maintain a constant proportionality during growth into adulthood.

This has been confirmed in a recent study in Nottingham (Patrick & Patel, 1986) in which children whose parents migrated from the Punjab to Britain have been compared with children of European stock. The two groups live side by side in deprived areas of the inner city and attend the same neighbourhood schools. Data for the girls are given in Figure 8. Also included are data reported by Bhattacharya and Banerjee (1966) for girls resident in the same region of North India: those girls might possibly have been the mothers of the Indian children studied in Nottingham. The data for boys, not shown here, demonstrate similar relationships. At the same age or at the same stature, the European children have larger ventilatory capacities than the Indian children in Britain, who in their turn have larger values than their counterparts in India. The 13% ethnic difference among European and Indian children is clear.

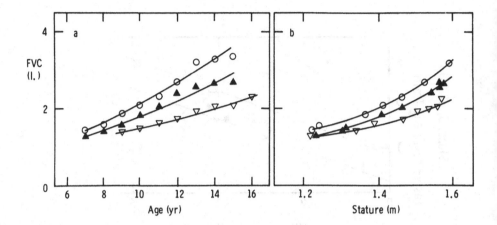

Figure 8:   FVC plotted against (a) age and (b) stature in three groups of girls.
              O, Europeans in UK; ▲, Indians in UK (data of Patrick & Patel); ▽
              Indians in India.   (Data of Bhattacharya & Banerjee.)

It is possible that some of the 5% to 15% difference between the curves
for the two sets of Indian children may be accounted for by methodological
differences: the observers, the instruments and the protocol were different,
though the posture was the same.   However, it is possible that some unidentified
environmental factor - probably nutritional because these British-born Indians
have a 6% weight-for-height advantage (Ulijaszek & Nicholl, 1984) - has altered
the shape of these children of migrants and given them larger chests for the
same stature.   Unfortunately, neither trunk nor thoracic dimensions were
measured in this study, so the ethnic and environmental influences on the shape
factor remains to be elucidated in a direct collaborative investigation.

OCCUPATIONAL EXPOSURE TO DUST

Although the respiratory effects of occupational exposure to dust have
been extensively studied in developed countries (cf. Morgan & Seaton, 1975),
rather little attention has been paid to such environmental hazards in the tropics
(Bouhuys, 1970) where control of working conditions may be less strict.   Roy
(1956) has reported on the incidence of pneumoconiosis in Indian coal-miners, and
Hearn (1968) has described the epidemiology of bagassosis in Trinidadian sugar-
workers of Indian and Negro stock.   The effects of bysinossis on ventilatory
capacity has been described for East African workers by Gilson et al (1962) and

for Indians by Kamet et al (1981). Schilling (1981) has recently reviewed the problems of bysinossis worldwide.

In the coal industry, for example, mining practices and dust control measures may differ materially from those in developed countries. Jain and Patrick (1981) have studied ventilatory capacity in miners in the only colliery in West Africa: at Enugu, Eastern Nigeria. The Vitalograph and Peak-flow meter were used to measure FVC, FEV_1 and PFR in 291 face-workers, 286 other underground workers with less exposure to dust, and 98 surface workers. All were asymptomatic adult males, though no radiographs were available. Values have been corrected to age 40 and stature 1.7 m using Cole's (1975) proportional method, i.e. multiplying by $(1.7/stature)^2$. Figure 9 shows the ventilatory capacity variables for the three occupational groups. The flow indices are significantly less in subjects with greater dust exposure, face-workers having a deficit of up to 10%. This corresponds to the findings in coal-mines in Britain (Muir, 1975) and America (Hankinson et al, 1977). However, the FVC is greatest in the face-workers while the FEV_1 shows no difference between groups. This suggests that the extra activity of the face-workers, particularly their work with shoulder girdle muscles, has trained the respiratory muscles to provide a greater FVC despite the demonstrable contrary effects of the dust. Figure 10 shows the ventilatory capacity variables, adjusted for stature where appropriate, plotted against age to show the rate of decline in this cross-sectional study. The slopes are the partial regression coefficients for age when stature has been allowed for (about 25 ml/year for FVC, see previous section). Those variables reflecting flow decline 15% more rapidly in face-workers than in the other subjects, but the FVC declines less quickly.

No effect of smoking was seen, though the small numbers of smokers and the relatively short duration of the smoking habit would have made it difficult to detect. So this colliery provided an example of high dust exposure but low smoking habit, a contrast to the picture in the mines in industrialised countries with controlled low dust exposures but more smoking (Patrick, 1981). In order to separate the effects of these two factors, the opportunity might be taken to compare such temperate and tropical working environments before the differences disappear.

SMOKING

Smoking provides a gross example of the deleterious effects of air pollution on ventilatory function. Many studies in developed countries have shown lower values for ventilatory capacity in current smokers compared with

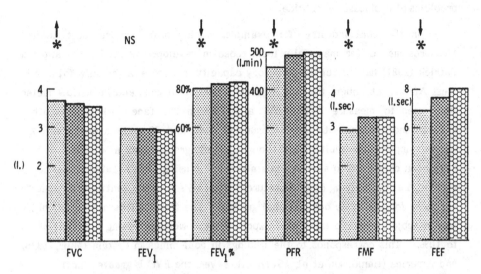

Figure 9:   Six indices of ventilatory capacity in three groups of Nigerian coal-
miners. ▦ face-workers; ▨ other underground workers;
▩ surface workers. The arrows indicate significant differences
between the face-workers and the others. (Data of Jain &
Patrick.)

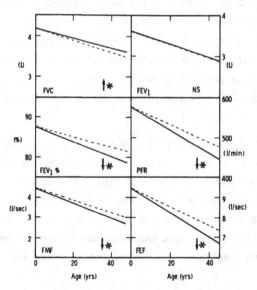

Figure 10:   The regression on age of six indices of ventilatory capacity
(adjusted to 1.7 m where appropriate) for (———) Nigerian coal-face
workers and (- - - -) other Nigerian colliery workers. The arrows
indicate significant differences between the face-workers and the
others. (Data of Jain & Patrick.)

non-smokers, and values which are even lower in those with the greater cumulative dose of smoke, measured in pack-years (Seltzer et al, 1974; Hubert et al, 1982). There is also a more rapid decline in ventilatory capacity in smokers studied longitudinally (Love & Miller, 1982). Tager et al (1983) have also shown that the growth of ventilatory function in children in adversely affected (to the extent of about 10% over five years) by the mother's smoking. In terms of its effect on working capacity, smoking has similarities with undernutrition: it increases time off work and reduces productivity. Few studies as yet, however, have established the impact of smoking on ventilatory and work capacity in tropical populations.

Reliable data on the current prevalence of smoking in tropical populations are hard to come by. Small studies suggest that in many tropical countries men are now as likely to be smokers as are men in developed countries, but there are wide regional variations (WHO, 1983) Hearn (1968) and Miller et al (1970) found that about half their subjects in Trinidad and Guyana were smokers, and Cotes et al (1974) and Anderson (1976) found that a large majority of their New Guinean subjects smoked either local or imported cigarettes. Among the Nigerian colliery workers studied by Jain and Patrick (1981) only 7.5% smoked cigrarettes while 80% of the remainder took their tobacco in the form of a locally-made snuff. However, it is noteworthy that 22% of the younger miners (under 35 years of age) smoked, compared with only 5% of the over-35s. This new habit is appealing to younger men who may continue with it for a lifetime.

An alternative source of data about smoking comes from the statistics of national imports and sales of cigarettes Taha and Ball (1983) have shown how these have increased dramatically in recent decades in several African countries, where there is no restriction on advertising and no government warning on the packet. They described the phenomenon as "the coming epidemic". However, there is little evidence that the effects of smoking on the respiratory system are yet appearing in tropical populations.

Smoking may be expected to take its toll in the tropics in the next decades as it has done already in industrial societies in temperate regions, unless some genetic or environmental factor means that cigarette smoke has a less deleterious effect in tropical populations. Seltzer et al in their large 1974 study of black subjects in North America found no difference in ventilatory capacity attributable to smoking, whereas there was a clear and significant smoking-related decrement among white smokers. In Jain and Patrick's (1981) study in Nigerian coal-miners, smoking had no influence on FVC when added as a separate term in the multiple regression. A formal large-scale study of the effect of smoking on pulmonary function in tropical populations is needed to answer this puzzling question.

We cannot leave the topic of smoking without broadening the enquiry to the wider issue of tobacco production in tropical countries, half of which is smoked in the country of origin.   The use of the resources of human labour and agricultural land for the cash cropping of tobacco, the destruction of forests to provide wood for the drying of tobacco leaves, and the displacement of food crops by tobacco are all important factors disturbing tropical ecosystems where food supplies for growing populations are marginal.   A multi-disciplinary study of all these aspects of the tobacco industry in a sample tropical location would fit exactly with one of the four primary thrusts of the programme outlined for the Decade of the Tropics by Solbrig and Golley (1984).   They point out that "there have been few collaborative studies between biologists and social scientists in which the economic, social, ethical and cosmological systems of human society are analysed in relation to the environmental - and more particularly biological - constraints and potentialities which surround it".   In developed countries the tobacco habit is at last on the decline. However, to the farmer, the transnational companies, the government revenue officers and even the consumer in the tropics, the short-term advantages of tobacco promotion may still appear to outweigh the wider and longer term disadvantages to the nutrition and health of the community.

CONCLUSIONS

Ventilatory capacity, generally measured by single-breath tests, is influenced by a range of constitutional and environmental factors.   The main ones, besides age and size, are ethnicity, customary physical activity and exposure to dust and smoke.   Although the operation of these factors has been demonstrated qualitatively in tropical populations, their relative importance has not been convincingly established, and much variance in ventilatory capacity remains to be explained.   In particular, the rates of decline attributable to age and to smoking need to be compared between different ethnic groups living in the tropics.   Furthermore, environmental influences, like those of physical activity and nutrition, on the shape and therefore the capacity of the thorax, have to be established.   This in turn will provide a better understanding of the variations in aerobic capacities and thus physical working capacities in tropical populations. Finally, the full impact of the tobacco industry on men and women living in a tropical ecosystem needs to be critically evaluated.

ACKNOWLEDGMENTS

This work could neither have been started nor continued without the help and support of Dr. J. E. Cotes. The hospitality and collaboration of colleagues and students in Nigeria, New Guinea, Sudan and Libya, as well as in Nottingham, are also gratefully acknowledged.

REFERENCES

Anderson, H.R. (1976). Respiratory abnormalities and ventilatory capacity in a Papua New Guinea island community. American Review of Respiratory Disease, **114**, 537-548.

Ballal, M.A., Fentem, P.H., Macdonald, I.A., Sukkar, M.Y. & Patrick, J.M. (1981). Physical condition in young adult Sudanese: a field study using a self-paced walking-test. Ergonomics, **25**, 1185-1196.

Beall, C.M. & Goldstein, M.C. (1987). Sociocultural influences on the working capacity of elderly Nepali men, In: K.J. Collins & D.F. Roberts (eds.), Capacity for Work in Tropical Populations. Cambridge University Press.

Beral, V. & Read, D.J.C. (1971). Insensitivity of respiratory centre to carbon dioxide in the Enga people of New Guinea. Lancet, ii, 1290-1294.

Bhattacharya, A.K. & Banerjee, S. (1966). Vital capacity in children and young adults of India. Indian Journal of Medical Research, **54**, 62-71.

Binder, R.E., Mitchell, C.A., Schoenberg, J.B. & Bouhuys, A. (1976). Lung function among black and white children. American Review of Respiratory Disease, **114**, 955-999.

Bouhuys, A. (1970). Pulmonary function measurements in epidemiological studies. Bulletin Europeen de Physiopathologie Respiratoire, **6**, 561-578.

Cole, T.J. (1975). Linear and proportional regression models in the prediction of ventilatory function. Journal of the Royal Statistical Society (Series A), **138**, 297-338.

Cotes, J.E. (1976). Genetic and environmental determinants of the physiological response to exercise, In: Medicine and Sport 9: Advances in Exercise Physiology, pp. 188-202. Basle: Karger.

Cotes, J.E. (1979). Lung Function (4th edn.). Oxford: Blackwell Scientific Publications.

Cotes, J.E. (1987). Is measurement of aerobic capacity a realistic objective? In: K.J. Collins & D.F. Roberts (eds.), Capacity for Work in Tropical Populations. Cambridge University Press.

Cotes. J.E. & Davies, C.T.M. (1969). Factors underlying the capacity for exercise: a study in physiological anthropometry. Proceedings of the Royal Society of Medicine, **62**, 620-624.

Cotes, J.E., Berry, G., Burkinshaw, L., Davies, C.T.M., Hall, A.M., Jones, P.R.M. & Knibbs, A.V. (1973). Cardiac frequency during submaximal exercise in young adults: relation to lean body mass, total body potassium and amount of leg muscle. Quarterly Journal of Experimental Physiology, **58**, 239-250.

Cotes, J.E., Anderson, H.R. & Patrick, J.M. (1974). Lung function and the response to exercise in New Guineans: role of genetic and environmental factors. Philosophical Transactions of the Royal Society, Series B, **268**, 349-361.

Cotes, J.E., Dabbs, J.M., Hall, A.M., Lakhera, S.C., Saunders, M.J. & Malhotra, M.S. (1975). Lung function of healthy young men in India: contributory roles of genetic and environmental factors. Proceedings of the Royal Society, Series B, **191**, 413-425.

Ghesquiere, J.L.A. (1972). Physical development of working capacity of Congolese. In: D.J.M. Vorster (ed.), Human Biology of Environmental Change, pp. 117-120. London: IBP/HA.

Ghesquiere, J.L.A. & Karnoven, M.J. (1979). Lung volume and working capacity of young and adult Shi and Havu of the Kivu lake district in Zaire. Annals of Human Biology, **6**, 289-290.

Gilson, J.C., Stott, H., Hopwood, B.E.C., Roach, S.A., KcKerrow, C.B. & Schilling, R.S.F. (1962). Byssinosis: the acute effect on ventilatory capacity of dusts in cotton ginneries, cotton, sisal and jute mills. British Journal of Industrial Medicine, **19**. 9-18.

Hankinson, J.L., Reger, R.B., Fairman, R.P., Lapp, N.L.& Morgan, W.K.C. (1977). Factors influencing expiratory flow rates in coal miners. In: W. H. Walton (ed.), Inihaled Particles, IV, pp. 737-755. New York: Pergamon.

Hearn, C.E.D. (1968). Bagassosis: an epidemiological, environmental and clinical survey. British Journal of Industrial Medicine, **25**, 267-282.

Hsi, B.P., Hsu, K.H.K. & Jenkins, D.E. (1983). Ventilatory functions of normal children and young adults: Mexican-American, white and black. III. Sitting height as a predictor. Journal of Pediatrics, **102**, 860-865.

Hsu, K.H.K., Jenkins, D.E., Hsi, B.P.. Bourhofer, E., Thompson, V., Tanakawa, N. & Hsieh, G.S.J. (1979). Ventilatory functions of normal children and young adults: Mexican-American, white and black, I. Spirometry. Journal of Pediatrics, **95**, 14-23.

Hubert, H.B., Fabsitz, R.R., Feinleib, M. & Gwinn, C. (1982). Genetic and environmental influences on pulmonary function in adult twins. American Review of Respiratory Disease, **125**, 409-415.

Jain, B.L. & Patrick, J.M. (1981). Ventilatory function in Nigerian coal-miners. British Journal of Industrial Medicine, **38**, 275-280.

Jones, P.R.M., Baber, F.M., Heywood, C. & Cotes, J.E. (1977). Ventilatory capacity in healthy Chinese children: relation to habitual activity. Annals of Human Biology, **4**, 155-162.

Kamat, S.R., Kamat, G.R., Salpekar, V.Y. & Lobo, E. (1981). Distinguishing byssinosis from chronic obstructive lung disease. American Review of Respiratory Disease, **124**, 31-40.

Levine, I., Evans, W.J., Winsmann, F.R. & Pandolf, K.B. (1982). Prolonged self-paced hard physical exercise comparing trained and untrained men. Ergonomics, **25**, 393-400.

Love, R.G. & Miller, B.G. (1982). Longitudinal study of lung function in coal miners. Thorax, **37**, 193-197.

Miller, G.J., Ashcroft, M.T., Swan, A.V. & Beadnell, H.M.S.G. (1970). Ethnic variation in forced expiratory volume and forced vital capacity of African and Indian adults in Guyana. American Review of Respiratory Disease, **102**, 979-981.

Miller, G.J., Cotes, J.E., Hall, A.M., Salvosa, C.B. & Ashworth, A. (1972). Lung function and exercise performance of healthy Caribbean men and women of African ethnic origin. Quarterly Journal of Experimental Physiology, **57**, 325-341.

Miller, G.J., Saunders, M.J., Gilson, R.J.C. & Ashcroft, M.T. (1977). Lung function in healthy boys and girls in Jamaica in relation to ethnic composition, test exercise performance and habitual physical activity. Thorax, **32**, 486-496.

Morgan, W.K.C. & Seaton, A. (1975). Occupational Lung Diseases. Philadelphia: Saunders.

Muir, D.C.F. (1975). Pulmonary function in miners working in British collieries: epidemiological investigations by the National Coal Board. Bulletin Europeen de Physiopathologie Respiratoire, **11**, 403-414.

Norgan, N.G., Ferro-Luzzi, A. & Durnin, J.G.V.A. (1974). The energy and nutrient intake and the energy expenditure of 204 New Guinean adults. Philosophical Transactions of the Royal Society, Series B, **268**, 309-348.

Patrick, J.M. (1976). Respiratory responses to CO_2-rebreathing in Nigerian men. Quarterly Journal of Experimental Physiology, **61**, 85-93.

Patrick, J.M. (1981). Smoking, coal, asbestos and the lungs. British Medical Journal, **283**, 675.

Patrick, J.M. & Cotes, J.E. (1974). Anthropometric and other factors affecting respiratory responses to carbon dioxide in New Guineans. Philosophical Transactions of the Royal Society, Series B, **268**, 363-373.

Patrick, J.M. & Cotes, J.E. (1978). Cardiac output during submaximal exercise in New Guineans: the relation with body size and habitat. Quarterly Journal of Experimental Physiology, **63**, 277-290.

Patrick, J.M., Bassey, E.J. & Fentem, P.H. (1983). The rising ventilatory cost of bicycle exercise in the seventh decade. Clinical Science, **65**, 521-526.

Patrick, J.M. & Patel, A. (1986). Ethnic differences in the growth of lung function in children: a cross-sectional study in inner-city Nottingham. Annals of Human Biology, **13**, 307-316.

Paul, R., Fletcher, G.H. & Addison, G. (1960). A comparative study between Europeans and Africans in the mining industry of Northern Rhodesia. Medical Proceedings, **6**, 69-74.

Roberts, D.F. (1987). Genetic factors in working capacity, In: K.J. Collins & D.F. Roberts (eds.), Capacity for Work in Tropical Populations. Cambridge University Press.

Rossiter, C.E. & Weil, H. (1974). Ethnic differences in lung function: evidence for proportional differences. International Journal of Epidemiology, **3**, 55-61.

Roy, K.B. (1956). Pneumoconiosis in central Indian coal-mines. British Journal of Industrial Medicine, **13**, 184-186.

Schilling, R.S.F. (1981). Worldwide problems of byssinosis. Chest, **79**, 3-6S.

Schoenberg, J.B., Beck, G.J. & Bouhuys, A. (1978). Growth and decay of pulmonary function in healthy blacks and whites. Respiration Physiology, **33**, 367-393.

Seltzer, C.C., Siegelaub, A.B., Friedman, G.D. & Collen, M.F. (1974). Differences in pulmonary function related to smoking habits and race. American Review of Respiratory Disease, **110**, 598-608.

Solbrig, O.T. & Golley, F.B. (1984).   A decade of the tropics.   Biology International, Special Issue **2**.

Tager, I.B., Weiss, S.T., Munoz, A., Rosner, B. & Speizer, F.E. (1983). Longitudinal study of the effects of maternal smoking on pulmonary function in children. New England Journal of Medicine, **309**, 699-703.

Taha, A. & Ball, K. (1980).   Smoking and Africa: the coming epidemic.   British Medical Journal, **280**, 991-993.

Tanner, J.M. (1949).   Fallacy of per-weight and per-surface area standards and their relation to spurious correlation.   Journal of Applied Physiology, **2**, 1-15.

Tanner, J.M. (1964).   The Physique of the Olymphic Athlete.   London: Allen & Unwin.

Ulijaszek, S.J. & Nicholl, A. (1984).   Weights, heights and skinfold thickness of multi-ethnic inner city children.   Annals of Human Biology, **11**, 471.

Weiner, J.B. & Lourie, J.A. (1969).   Human Biology: A Guide to Field Methods. Oxford: Blackwell Scientific Publications.

World Health Organization (1983).   Smoking control strategies in developing countries.   WHO Technical Report Series No. **695**.

IS MEASUREMENT OF AEROBIC CAPACITY A REALISTIC OBJECTIVE?

J. E. COTES

University Department of Occupational Health,
University of Newcastle upon Tyne, U.K.

INTRODUCTION

In many communities and occupations, human muscle is still the main source of power for the performance of physical work. The power output is determined by the quantity of muscle and the effectiveness with which it is used; this in turn is influenced by the adequacy of nutrition, the extent of any disease and the levels of physical training and skill. A high absolute power output is of benefit to the community in question, a high output relative to the quantity of muscle is evidence for optimal efficiency in the individuals concerned. Power output for work of a few minutes' duration is reflected by the maximal oxygen uptake (aerobic capacity). In physiological terms it is a measure of the combined function of all components of the oxygen transport chain from respired air to muscle mitochondria. In ergonomic terms it is the maximal power output in the circumstances of the test which may be on one or more of a treadmill, cycle ergometer or step. These forms of exercise utilise the main muscle mass of the body but in different ways and so the result which is obtained varies with the type of ergometer. It is also influenced by environmental factors including ambient temperature, but in the laboratory this can usually be controlled.

MAXIMAL OXYGEN UPTAKE IN POPULATION STUDIES

The procedure for measuring maximal oxygen uptake entails the subject exercising at progressively increasing intensity up to the maximum. In objective terms this is the point at which oxygen uptake reaches a plateau and does not increase with a further increment in external work. Subjectively, maximal exercise is associated with unsustainable breathlessness, palpitation, fatigue or other symptom and impending collapse. The attainment of maximal exercise as defined requires some competence at the exercise in question, considerable dedication or a willingness to be coerced, and absence of atherosclerosis which might otherwise cause angina, intermittent claudication or significant electro-cardiographic abnormality.   Persons who meet these criteria can achieve a reproducible maximal oxygen uptake; they include a high proportion of male children and adolescents, athletes, persons whose employment requires a high level of physical fitness, and those members of developing communities who have experience of the type of exercise which is to be used.   However, persons capable of attaining an acceptable maximal oxygen uptake are usually in the minority in population studies.

Here are four examples: (1) Amongst New Guineans on Karkar Island and at Lufa there was a high degree of cooperation and willingness to undertake test exercise which was in an air-conditioned laboratory;   however, a large proportion of subjects failed to meet the objective criterion for maximal exercise because they had never cycled before (Cotes, Anderson & Patrick, 1974).   (2) Soldiers in the Indian Army performed two periods of progressive exercise up to the symptom-limited maximum.   Their cooperation was superb but few met the objective criterion due to the high temperature and humidity of the laboratory (Cotes, 1976).   (3) Shipyard workers aged 23 to 47 years volunteered for a test of maximal exercise as part of an occupational survey.   In the event, 15% declined to exercise, 16% did not exert themselves to meet the objective criterion or the lesser one of maximal heart rate that was within 10 beats per minute of the reference value.   A further 15% developed notable electrocardiographic abnormalities during the test, 9% had other difficulties, and only 45% achieved the target (Weller et al, 1985).   (4) Amongst mine-rescue workers and divers whose employment depended on their having an adequate exercise capacity, a high proportion met the objective criterion (Robertshaw et al, 1984;   Cotes, 1987).   Thus a test of maximal exercise is practicable in appropriate circumstances.   It is useless for an epidemiological study where in order to draw representative conclusions results must be available for at least 95% of the subjects (Oldham, personal communication).

INTERPRETING MAXIMAL OXYGEN UPTAKE

Maximal oxygen uptake ($\dot{n}O_2$ max) (units in mmol/min, cf. $\dot{V}O_2$ max in l/min) reflects the maximal power output which in engineering is often expressed as power to weight ratio.   The physiological equivalent is $\dot{n}O_2$ max per kg body mass or fat-free mass.   However, this mode of expression implies that $\dot{n}O_2$ max is proportional to mass which is not the case.   Instead, the relationship is a linear regression with a significant constant term.   The existence of this term inevitably leads to $\dot{n}O_2$ max/mass being negatively correlated with mass;   the

Table 1:   Linear regression equation of $\dot{n}O_2$ max on body mass and fat-free mass

---

$\dot{n}O_2$ max (mmol/min) = 1.58 fat-free mass (kg) + 34.88 S.E.E. = 20.04

$\dot{n}O_2$ max (mmol/min) = 0.57 body mass (kg) + 88.92 S.E.E. = 21.74

n = 156: $\bar{y}$ = 130.49

---

relationship is an arithmetic artefact.   The error was reported in detail by Tanner (1949) and is illustrated in the present context by the results for the shipyard workers referred to above (Table 1).   The same error invalidates the use of ventilation equivalent (ventilation ÷ O_2 uptake) (Lloyd & Patrick, 1963), and oxygen pulse (oxygen uptake ÷ cardiac frequency).

Since $\dot{n}O_2$/mass is not a valid index there is need for an alternative way of comparing power output in subjects of different mass.   The power output at the mean mass of the group, calculated from the linear regression, can be used. Alternatively, the result can be expressed per unit of body muscle or body cell mass;   the latter index may be based on measurement of total body potassium which in one study was proportional to maximal oxygen uptake for both male subjects alone and subjects of both sexes combined (Cotes & Davies, 1969).

ESTIMATING MAXIMAL OXYGEN UPTAKE

As an alternative to direct measurement, maximal oxygen uptake was formerly obtained from the relationship of cardiac frequency to oxygen uptake during submaximal exercise;   this was extrapolated to the estimated maximal cardiac frequency using the reference values of Åstrand and Rodahl (1977). However, neither the relationship nor the reference value is exact and the variability leads to the derived maximal oxygen uptake having wide confidence limits.   The alternative use of related variables to estimate maximal oxygen uptake was explored preparatory to the International Biological Programme (Cotes et al, 1969). The following prediction equation was then obtained:

$$\dot{n}O_2 \text{ max (mmol/min)} = 8.48 \text{ FFM(kg)} + 7.14 \text{ FVC(l)}$$
$$- 0.49 \text{ fC}_{67} \text{ (min)} + 61.6 \text{ (SD 11.6)}$$

where FFM is fat-free mass, FVC is forced vital capacity, and fC_{67} is cardiac frequency at an oxygen uptake of 67 mmol/min.

The relationship gave a less good prediction in other populations.   It has

Table 2:   Regression equation for predicting $\dot{n}O_2$ max from $\dot{n}O_2$ at $R_{1.0}$ and other variables (n = 156)

---

nO_2 max (mmol/min) = 71.45
$\qquad$ + 0.645 $\dot{n}O_2$ at $R_{1.0}$ (mmol/min)
$\qquad$ - 0.287 fC_{45} (min) *
$\qquad$ + 0.804 fat-free mass (kg)
$\qquad$ - 0.743 % fat

$$\bar{y} = 130.49 \text{ mmol/min}$$
$$\text{S.E.E.} = 11.96$$
$$\text{Coefficient of variation} = 9.0\%$$

*Cardiac frequency at $\dot{n}O_2$ = 45 mmol/min

---

since been further developed by inclusion of oxygen uptake at a respiratory exchange ratio of unity ($\dot{n}O_2$ at $R_{1.0}$) (Table 2). In this form the equation has provided accurate prediction ($\pm$ 9%) in two separate studies in the U.K. (Mortimore & Reed, 1982; Weller et al, 1985). Its usefulness in other situations has still to be established.

VALIDITY OF ESTIMATED $\dot{n}O_2$ max

The $\dot{n}O_2$ max is used to characterise an individual or group, to register changes associated with an alteration in habitual activity or other variable and to throw light on the underlying factors. For this purpose the $\dot{n}O_2$ max is used as the dependent variable in a multiple regression analysis; the independent variables may include age, the score for habitual activity, whether or not the subject smokes, and other variables. In the case of the shipyard workers, the estimated $\dot{n}O_2$ max was a good approximation to the directly measured value. There is preliminary evidence for it providing a good estimate of changes associated with physical training. For analysis of underlying factors the estimated $\dot{n}O_2$ max when compared with the directly measured value was found to be similarly related to age and to the score for habitual activity. However, whereas $\dot{n}O_2$ max was significantly related to smoking this was not the case for estimated $\dot{n}O_2$ max. This difference was due to smoking affecting the $\dot{n}O_2$ at $R_{1.0}$ which was one of the terms used in the estimation. Thus in this instance the transformation from $\dot{n}O_2$ at $R_{1.0}$ and other component variables to estimated $\dot{n}O_2$ max entailed loss of information and analysis of the component variables was more informative.

CONCLUSIONS

For an index to be of use in a population study it should ideally be capable of registration on 100% of subjects. A lapse rate of more than 5% is likely to invalidate the findings. This consideration excludes maximal oxygen uptake; $\dot{n}O_2$ max is only of use for investigating subsamples selected for their ability to complete the test. In such individuals the result should be reported with due regard to arithmetic probity and not per kg body mass. $\dot{n}O_2$ max can be estimated from indices of submaximal exercise including $\dot{n}O_2$ at $R_{1.0}$: the estimate is suitable for making comparisons between groups and for investigating the role of habitual activity. However, for investigating the factors which contribute to exercise performance, greater accuracy is achieved by analysing the submaximal indices themselves. Of these the $\dot{n}O_2$ at $R_{1.0}$ contains the most information but indices recommended previously including ventilation and cardiac frequency at an oxygen uptake of 45 mmol/min (1.0 l/min) are also informative (Cotes et al, 1969) and have the advantage of being attainable by almost all members of normal adult populations. There is a strong case for their being included in studies of populations carried out under the programme for the Decade of the Tropics. Measurement of maximal oxygen uptake should be confined to selected subsamples and the selection taken into account when the results are reported.

REFERENCES

Åstrand, P.O., Rodahl, K. (1977). Textbook of Work Physiology, 2nd edn, pp. 333-365. London: McGraw-Hill.

Cotes, J.E. (1976). Genetic and environmental determinants of the physiological response to exercise. Medicine and Sport, vol. 9: Advances in Exercise Physiology, pp. 178-202. Basel: Karger.

Cotes, J.E. (1987). Exercise fitness of divers, In: T. G. Shields (ed.), Fitness to Dive. Proceedings of Diving Symposium. Institute of Environmental and Offshore Medicine, Aberdeen. (In press.)

Cotes, J.E., Anderson, H.R. & Patrick, J.M. (1974). Lung function and the response to exercise in New Guineans: role of genetic and environmental factors. Phil.Trans.Roy.Soc.Lond.B, **268**, 349-361.

Cotes, J.E. & Davies, C.T.M. (1969). Factors underlying the capacity for exercise: a study in physiological anthropometry. Proceedings of the Royal Society of Medicine, **62(6)**, 620-624.

Cotes, J.E., Davies, C.T.M., Edholm, O.G., Healy, M.J.R. & Tanner, J.M. (1969). Factors relating to the aerobic capacity of 46 healthy British males and females, aged 18 to 28 years. Proceedings of the Royal Society of London (Series B), **174**, 91-114.

Lloyd, B.B. & Patrick, J.M. (1963). Respiration and metabolic rate during active and passive exercise in man. Journal of Physiology, **165**, 67.

Mortimore, I.L. & Reed, J.W. (1982). The effect of diet on the respiratory exchange ratio and aerobic capacity in man. Journal of Physiology, **330**, 59-60.

Robertshaw, S.A., Reed, J.W., Mortimore, I.L. & Cotes, J.E. (1984). Submaximal alternatives to the Harvard pack index as guides to maximal oxygen uptake (physical fitness). Ergonomics, **27**, 177-185.

Tanner, J.M. (1949). Fallacy of per-weight and per-surface area standards and their relation to spurious correlation. Journal of Applied Physiology, **2**, 1-15.

Weller, J.J., El-Gamal, F.M., Parker, L., Reed, J.W., Bridges, N.G., Chinn, D.J. & Cotes, J.E. (1985). Estimating the capacity for exercise of shipyard workers. Clinical Science, **68** (Suppl. 11), 45.

FUNCTIONAL CONSEQUENCES OF MALNUTRITION

MALNUTRITION, WORK OUTPUT AND ENERGY NEEDS

R. MARTORELL and G. ARROYAVE

*Food Research Institute, Stanford University,
Stanford, California, U.S.A.*

INTRODUCTION

There continues to be a great deal of interest, particularly among nutritionists and development economists, in the functional significance of malnutrition. The term "malnutrition" generally refers to all deviations from adequate nutrition, including energy undernutrition or overnutrition and specific deficiencies or excesses of essential nutrients. In the context of this presentation, however, the term "malnutrition" refers to the nutritional situation characteristic of low socioeconomic populations in underdeveloped areas of the world. The diets in these populations are characterised by two predominant features: (1) insufficiency in amounts which results in a deficit of food energy for the majority of the people; and (2) limited quality and inadequate number and variety of food items, aspects which constrain the achievement of a balanced diet with appropriate concentrations of all the essential nutrients (nutrient density). The consequence is the concurrence of undernutrition and specific nutritional deficiencies. Alan Berg (1981), from the World Bank, sees malnutrition not only as an outcome of poverty but as a deterrent to economic progress. One way through which Berg (1981) believes malnutrition is damaging is by limiting work capacity and productivity in workers. At the same time, there are those that have criticised nutritionists for exaggerating the extent of world hunger and malnutrition (Seckler, 1980; Poleman, 1981; Sukhatme & Margen, 1982). In their view, malnutrition is not as common or as serious as is generally claimed and therefore should not be included among the significant factors limiting productivity of workers.

It is with these ideas as background that we will attempt to outline the conceptual linkages between malnutrition and productivity, to review the evidence supporting these linkages, and to address the general question of the role of malnutrition in productivity.

CONCEPTUAL FRAMEWORK

In considering the effects of malnutrition on work capacity and work output in adults, long-term developmental effects must be distinguished from limitations imposed by current nutritional problems.   This is the intent of Figure 1 which guides much of our discussion.   We indicate that malnutrition during the rapid stages of development in pregnancy and early childhood will retard physical growth and lead to a reduced adult lean body mass and through this mechanism directly affect work capacity as measured by maximal oxygen uptake (VO_2 max).   Work capacity and output are undoubtedly affected by behavioral characteristics such as intelligence and motivation, aspects which in turn may be adversely influenced by malnutrition in childhood.   For these reasons we speculate that malnutrition may influence work capacity and work output through behavioral as well as physical mechanisms.

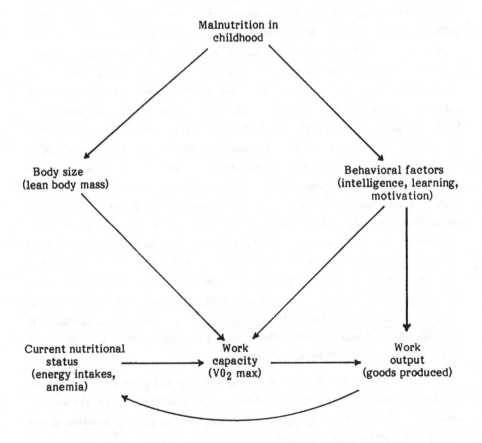

Figure 1: General linkages between nutrition and work output

A poor current nutritional status definitely places a ceiling on the capacity to do work. The two nutritional problems which have been most intensively studied in developing countries are low energy intakes and anaemia. Figure 1 also implies that work capacity affects work output, a linkage certainly true in strenuous and moderate levels of activity but not necessarily the case in light activities. Finally, Figure 1 shows that the level of work output determines income and this in turn affects current dietary intakes and nutritional status.

MALNUTRITION AND BODY SIZE

It is well known from animal and human studies that malnutrition during the developmental phase will result in adults of small body size. In humans, there is evidence indicating that the period of pregnancy and the first three to five years are the times during which size is greatly affected. Studies indicate that by the time children in developing countries reach five years of age, the damage will have already been done. By five years of age, many of these children will already be destined to be short adults (Martorell & Habicht, 1986).

This observation should come as no surprise to anyone. Children are highly vulnerable to nutritional problems during infancy and the weaning period. While nutritional needs relative to body weight are greater than at any time thereafter, nutrient intakes are in fact at their worst. Mothers may be malnourished and unable to produce adequate amounts of breast milk to sustain normal growth in their infants beyond three to six months of life. Matters are compounded by the nutritionally inadequate complementary feedings usually given to infants. In addition, children at these young ages are highly susceptible to infections, particularly diarrhoeal diseases, problems which adversely affect nutritional status in several ways (Chen & Scrimshaw, 1983). Diarrhoeas and other problems appear to reduce dietary intakes in children and sometimes mothers may also purposely restrict the sick child's diet. Nutrients are diverted from growth by the catabolic processes that accompany infections, and the rapid intestinal transit and malabsorption occurring in diarrhoea may also limit nutrient availability. Additional common problems are vomiting and fevers, both with obvious nutritional implications. The end result of poor diets and infection, of course, is that children grow very poorly at a time when they should be growing very fast.

There are some studies, such as the one by Satyanarayana et al (1980) from India which demonstrate clearly the importance of the early years for final size. In a unique study, heights of boys from rural Hyderabad were measured at

Table 1   Height at 5.0 and 17.5 years of age in rural Indian children*

| Groups** | No. | Initial point (5 yrs) | | Current point (17.5 yrs) | | Growth between (5 & 17.5 yrs) |
		Age (yrs: 1965)	Height (cm)	Age (yrs: 1978)	Height (cm)	(>150 months) (cm)
I	23	5.04	105.0	17.43	164.5	60.5
II	25	5.22	99.8	17.45	160.1	61.6
III	30	5.14	95.6	17.41	157.0	62.6
IV	14	5.16	88.5	17.39	149.0	62.2
Pooled S.D.	14	0.32	3.21	0.28	6.40	6.23
F ratio		1.16	85.45	0.30	16.12	0.47
Signifi- cance level		>0.05	<0.001	>0.05	<0.001	>0.05

* From Satyanarayana et al, 1980.

** Groups were defined on the basis of height at 5 years of age and in reference to the Boston median and standard deviation as follows: Group I, values within the range 0 to -2 S.D.; Group II, -2 S.D. to -3 S.D.; Group III, -3 S.D. to -4 S.D.; and Group IV, less than -4 S.D. Mean z-scores (based on NCHS reference data) were -1.88, -2.55, -3.03, and -4.26, respectively at age 5 and -1.07, -2.20, -3.11, and -4.65, respectively at 17.5 years of age.

seventeen years of age and related to their heights at five years of age.   The sample was divided into four groups according to the level of stunting at age five (see footnote, Table 1).   The level of stunting was considerable in all groups.   All groups gained about the same amount in height from five to seventeen years and the average increase, 62 cm, was only 5 cm less than that of children from the United States (i.e., NCHS growth charts).   Therefore, height deficits observed at age five continue unchanged into adolescence and the relative rank between groups is maintained: all remain short but the shortest and the tallest groups at five are still the shortest and the tallest respectively at seventeen years.

There has been a great deal written about ethnic differences in genetic potential for growth (Martorell & Habicht, 1986).   Studies increasingly point out that there are few differences in body size between young children of different ethnic groups if they are adequately fed and healthy. There are indications from India (Rao & Sastry, 1977) and Guatemala (Johnston et al, 1973) that ethnic differentiation may occur largely during the adolescent period.   But often, genetically caused differences in adult body size and body proportions turn out to be less than many had anticipated. A case in point is the Japanese population. Tanner and colleagues (1982) tell us that the secular trend toward greater heights will soon cease and that at this point, Japanese adults will only be 5-6 cm shorter than the British, a difference equivalent to one standard deviation. Also, an interesting point is that at this juncture, British and Japanese adults will have similar body proportions.

MALNUTRITION AND BEHAVIOUR

As pointed out by numerous authors, studies of the relationship between malnutrition and mental development are complicated by the fact that poor socioeconomic environments and malnutrition are intimately related (Pollitt, 1980; Brozek & Schurch, 1984). Analytically, it becomes difficult to separate the effects of poverty and malnutrition on mental development.

There is no doubt that long-term severe malnutrition, and particularly marasmus, is clearly associated with serious deficits in cognitive performance (Pollitt, 1980). The question regarding the effects of mild to moderate malnutrition remains controversial (Brozek & Schurch, 1984). Nutritional interventions have resulted in a range of improvements in cognitive tests but the magnitude of these effects have been small (Ricciuti, 1981). Small body size and poor performance in tests of mental development and poor attendance and low achievement in school are indeed correlated in children, but here the problem, as mentioned above, is establishing how much of this association is explained by the poorer social environment in which smaller children develop (Pollitt, 1980).

Much remains to be discovered about the mechanisms through which malnutrition affects mental development. Malnutrition may affect cognitive development directly by interfering in some fashion with brain development. Indirect mechanisms are also possible. Low energy intakes and anaemia, for example, may prevent children from a normal life of play and exploration of the environment. Similar nutritional problems in adults may limit the degree of parent-child interactions with detrimental effects for cognitive development. We know little about physical activity levels in children in developing countries and little about how such variation may affect development.

The hypothesis has been proposed that malnutrition in childhood, in combination with poor social environments, reduces intelligence and motivation, aspects which in turn affect work capacity and work output (Selowsky & Taylor, 1973). There is no evidence from developing countries to support or disprove a link with work capacity and productivity.

BODY SIZE, WORK CAPACITY, AND WORK OUTPUT

Calloway (1982) notes that "All things being equal, persons of small stature should have a clear advantage in resource limited societies" (p. 738). She goes on to say that "From this perspective, the limitation of linear growth (stunting) can be viewed as an adaptative response in that adapted individuals can make do with less food throughout adulthood, when total physiological demands are highest" (p. 738). However, and as Calloway notes, there are losses of function in other areas. Martorell (1985) has argued that in order to end up with small adults, a generation of children have to be subjected to inadequate diets and frequent infections. Many of the weakest do not survive. As we have noted, mental development will be affected, particularly if malnutrition is severe.

Growth retardation carries, therefore, a heavy price. But leaving this aside, we can ask whether being a small adult matters. Seckler (1980) would tell us not. He has proposed the "small but healthy" hypothesis and sees small body size as a desirable, true adaptation for countries like India. But, we know this is not true in terms of the capacity of women to bear and deliver healthy, well-nourished infants who are capable of surviving the stormy few years of life in developing countries (Martorell et al, 1981). More relevant to this meeting, we can ask whether body size makes a difference in terms of work capacity and work output.

Spurr (1983) has made it clear that stature and lean body mass are correlated. Maximal oxygen uptake (VO_2 max), the "gold standard" of cardio-respiratory fitness, is related to lean body mass. Taller people generally have larger body masses and therefore have larger maximal oxygen uptakes and hence higher work capacities. At a strenuous work load, those with a lower work capacity would be closer to a maximal effort and would function at faster heart rates (Spurr, 1983). Such overtaxed individuals may not be able to maintain the work pace for very long and may produce less. Khosla (1968) has made the interesting observation that many Olympic events such as short- and middle-distance running, jumping, and throwing are unfair for short people. Tall athletes have an unfair advantage because of their longer stride and greater lung capacities. Khosla (1968) argues that these events should be contested within height ranges because "Short champion throwers, runners, hurdlers and jumpers are waiting to be discovered" (p. 113).

Before turning to the literature on the relationship between body size and work capacity and productivity, we would like to emphasise two points. First, weight is not a useful measure of body size for our purposes since we are trying to isolate long-term nutritional effects from current nutritional problems. The anthropometric measures of choice instead would seem to be height or other measures of linear growth as indicators of long-term nutritional effects and weight for height or similar measures as indicators of current nutritional status. Second, we need to assess the significance of body size in terms of not just work capacity but work output. Jobs that require very low efforts but involve climbing may in fact be carried out more efficiently by small people. For instance, small Guatemalan Indians planting corn with a stick on very steep hills may be able to seed a field at a lower energy cost than if they were larger.

A number of studies have been carried out about body size and sugar cane cutting, a strenuous, energy-demanding job. Davies (1973) divided sugar cane cutters in Tanzania into high, medium and low producers based on the daily amount cut. The groups were similar with respect to height, weight, summed skinfolds, LBM, leg volume, and the circumferences of biceps and calf but in men younger than 35 years, correlations between daily output and weight, lean body mass and leg volume were statistically significant. Weight for height was not studied. Davies (1973) also reported that the total amount of sugar cane cut during the season was not significantly related to VO_2 max. However, VO_2 max was significantly related to the rate of work, that is, to the amount of cane cut per day. VO_2 max was also negatively associated with voluntary absences from work. In other words, fitter individuals were able to produce a given output in less time, and were less likely to take unpaid days to rest. The study implies that fitter individuals were employed for fewer days but does not comment on this aspect. Spurr et al (1977) divided Colombian cutters into good, average, and poor workers. The groups were different with respect to height, weight, and LBM but not percent body fat. A stepwise regression of the predictors of the daily tonnage of cane cut included three significant variables: VO_2 max, percentage body fat, and height. Weight for height was not studied by Spurr et al (1975).

Productivity in sugar cane workers has also been studied in Guatemala by Immink et al (1982). Regression analysis showed that upper-arm muscle area/height and percentage weight/height were not significantly related to productivity indicators. Tall workers were significantly more productive than short workers but the difference was small. However, the total effect of height was actually greater when account was taken of the fact that height was positively correlated with the number of days worked per week. This Guatemalan study shows, therefore, that taller sugar cane workers were somewhat more productive and more likely to show up for work.

Table 2:   Correlations between productivity, weight, and height in Indian
industrial workers, n = 46*

	1. Production	2. Weight	3. Height
1. Production	-	0.72 **	0.43 ***
2. Weight	-	-	0.51 **

Partial correlations: r 13.2 = 0.11 N.S.
r 12.3 = 0.65 **

```
     *      From Satyanarayana et al, 1977
    **      p < .0001
   ***      p < .01
```

A Jamaican study found that cane cutters whose weight for height was below 85 percent of the standard, cut significantly less cane than those whose weight for height was above 95 percent (Heywood, 1974).   Similarly, the productivity of road workers engaged in three activities - wheelbarrow work, ditch digging, and earth excavation - in Kenya was found to be significantly related to weight for height (Brooks et al, 1979).

A study from India (Satyanarayana et al, 1977) has looked at the relationship between body size and industrial factory work of less intensity than sugar cane cutting.   Specifically, the men studied made detonator fuses. Correlations between productivity, weight, height, and lean body weight from this study are shown in Table 2.   Height had a correlation of 0.43 with production, and weight had a correlation of 0.72 with production..   Both correlations were highly significant but the one involving weight was larger. The partial correlations are interesting.   The correlation between production and height, after adjusting for weight, is 0.11 and is not statistically significant. The correlation between weight and production, holding height constant, is 0.65 and highly significant.   Weight for height rather than stature would seem to be important for productivity in this setting.

There have been few studies of body size and work ouptut in farm households.   One study we are aware of is an econometric analysis which assessed the importance of weight for height and of height for agricultural and farm productivity in rural South India (Deolalikar, 1984).   The author found that weight for height was a significant predictor of both daily agricultural wages and on-farm labour productivity.   Height played no role at all.   It was estimated that an increase in weight for height from the sample mean of 86 percent to the normal value of 100 percent would result in an increase of 0.42 Rs or 17 percent of the daily wage rate of 2.45 Rs.

In summary, what generalisations can we offer on the basis of the studies relating body size and productivity? First, many of the work activities studied to date have been very strenuous ones such as sugar cane cutting. We know much less about effects on less demanding agricultural or industrial tasks. Second, unfortunately, not all studies have separated weight for height from height effects. Third, height appears to play a role only for sugar cane cutters in Latin America. Fourth, a number of studies indicate that weight for height is often an important predictor of productivity, particularly in less demanding work tasks.

ENERGY SUPPLEMENTS, ANAEMIA, BODY COMPOSITION AND WORK OUTPUT

It is well known that severe nutritional deficiencies and excesses seriously limit work performance and social function. Viteri (1982) expresses the view that these are unusual situations that should be dealt with because they threaten life, the issue of work performance under these circumstances being a secondary matter.

(a)   *Energy and work output*

Experience in industrial workers in Germany in World War II clearly indicate the relationship between energy intake and work output (Kraut & Muller, 1946). Raising daily caloric intake by about 500 kcal in 20 workmen improved the weight of material moved per hour from 1.5 tons to 2.5 tons in one setting while 31 coal miners whose rations increased from 2800 to 3200 kcal/day increased production from 7 to 9.6 tons per day. A far more convincing demonstration is the close correspondence in the trends of the energy provided by the rations and the output of the workers in the pits of the Ruhr coal mines from September 1939 to September 1944. So convinced were these German researchers of the importance of an adequate diet for production that in reference to the future of Germany, as they saw it in 1946, they ended their article with the phrase "Reconstruction is a problem of calories" (p. 497).

In a recent econometric study, it was found that current nutritional status of farm labourers in Sierra Leone, as measured by annual availability of energy, was associated with increased farm output, holding other inputs constant (Strauss, 1984). It must be pointed out that household-level data were used and that dietary measures of the actual energy intakes of individuals were not available for study.

One way by which we can test the importance of current nutritional status for work output is by means of food supplements with energy and iron. Among the available studies is a short-term supplementation trial carried out in Kenyan road workers (Wolgemuth et al, 1982). Workers were divided into two groups. One group received a supplement providing 200 kcal/day and the other a supplement giving 1000 kcal/day. Pre-intervention dietary intakes were 1350 kcal/day in women and 2086 kcal/day in men. The feeding period was four months. Though 224 workers participated in the study, baseline and final productivity measurements were available for only 47 cases. Being a member of the high-energy group was associated with a 12.5 percent improvement in productivity (p <.10). One could argue that the comparison is unfair in the sense that both groups compared were supplemented. This study had serious data collection problems and, at best, the research can only be said to suggest that large energy supplements have a modest effect on productivity when compared to small supplements. A short-term study from India (Satyanarayana et al, 1972) did not reveal productivity gains in coal workers receiving calorie supplements. This study is difficult to interpret because there was a major coal car constraint which may have limited the gains in productivity.

Viteri (1974) has found that caloric supplementation is positively asso-ciated with increased levels of energy expenditure and increased LBM and muscularity in a three-year study of agricultural workers in Guatemala. Also, time and motion studies revealed that the high-energy-intake group spent a significantly larger portion of the day working and was able to perform standardised tasks in significantly less time than a control group not receiving supplementation. It was concluded that this study demonstrates that insuffi-cient caloric intake imposes a ceiling on the total energy expenditure in agricultural labouring populations. An interesting qualitative finding of the study was that the unsupplemented workers were unable to continue to be active after work, usually taking long naps in the afternoon.

A larger study that included measures of work output was subsequently carried out by Viteri and coworkers (Immink & Viteri, 1981; Immink et al, 1984). The study took place as before in Guatemala. Workers from one community received 550 kcal/day as a liquid supplement (n = 95) while workers from a second community served as controls (n = 63) and also received a drink but with a negligible amount of calories. Pre-supplementation intakes were 3046 and 2899 kcal/day for experimental and control subjects (Immink & Viteri, 1981). The study lasted 30 months but only the data for the first fifteen months have been extensively presented. The investigators were able to show that energy intakes increased by about 280 kcal in the supplemented group indicating considerable

substitution of the supplement for normal food intake. These increased energy
intake levels did not appear to have produced a significant impact on work
output.   Body weights did not change so it was inferred that the better
supplemented subjects must have increased their level of off-work and leisure
activities.

(b)   *Anaemia and work output*

Anaemia has been shown by many to adversely affect work capacity.
Many have also shown it to be related to work output.   Popkin (1978)
demonstrated in multivariate analyses that the haemoglobin concentration was a
significant predictor of productivity with regard to the average daily output of
workers in loading, unloading, and tamping of soil in workers from the
Philippines. Anaemic workers were also found to be more frequently absent
from work than non-anaemic workers. Studies conducted in Indonesia (Basta et
al, 1979), Sri Lanka (Gardner et al, 1977), and Guatemala (Viteri & Torun, 1974)
have demonstrated that iron supplements to workers increase haemoglobin
concentrations and work output measured in terms of the Harvard step test or
the treadmill test. The studies in Indonesia (Basta et al, 1979) were interesting
in several respects. The design was a double-blind intervention trial. Tappers
and weeders in rubber plantations were included. Both treatment and placebo
groups received a daily incentive payment equivalent to about 5 to 7 percent of
minimum daily wages. The authors note that this sum was spent largely on food,
resulting in minor improvements in iron and in a significant amount of vitamin C.
The findings with respect to the latex workers were dramatic and are shown in
Figure 2. The values are given as a percentage of the latex production of non-
anaemic subjects in the baseline phase (= 100%).   As seen in the left, anaemic
subjects produced 81.3 percent of the amount produced by the non-anaemic
workers and this difference was statistically significant.   During the
intervention, all groups, whether anaemic or not, improved. Explaining positive
changes in all subjects is difficult.   One possibility is that payments and
participation in the investigation motivated all the men to work harder.
However, the investigation is still useful in that it shows that those who were
anaemic and who were treated with iron improved significantly more than those
anaemic workers who were given a placebo. For workers who were not anaemic,
the degree of improvement was independent of the type of pill given and thus the
data were combined by the authors as shown in Figure 2.

(c)   *Are supplements to workers effective in raising productivity?*

What can be said about the supplementation experiments? What do they
tell us about the role of current nutritional status as a factor in productivity?

Figure 2:   Changes* in latex production in Indonesian workers**

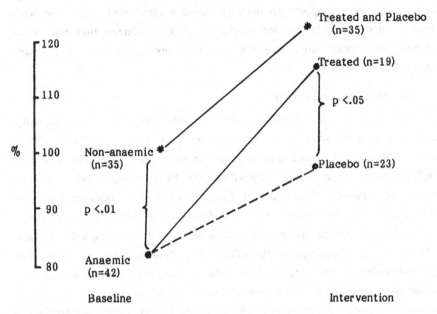

* Values are expressed as a percent of the latex output of non-
  anaemic (>13 g Hb) subjects in the baseline phase. The actual
  values for this group are 25.77 ± 9.55 kg of latex/day.

** Data from Basta et al, 1979.

First, there have been few large-scale intervention trials, particularly for
energy. Second, the studies by Viteri and colleagues in Guatemala dealing with
energy supplement have not revealed large effects on productivity. They suggest
that non-leisure activities may have been greatly affected but these issues have
not been properly documented.   Perhaps the energy intakes of the Guatemalan
workers studied by Viteri were not very deficient.   Energy supplements to
populations with high energy needs (Kraut & Muller, 1946) or low intakes
(Walgemuth et al, 1982) suggest larger effects on productivity.   Third, the
evidence is clear for iron deficiency anaemia: its correction leads to increased
productivity.   From the design and methodology point of view, it is much easier
to carry out a study dealing with iron deficiency anaemia than with energy.   A
case in point is the study by Basta et al (1979) which despite some complications
could still be interpreted unequivocally because of the power of the design of a
double-blind intervention trial.

ENERGY NEEDS AND ENERGY BALANCE

Implicit in our discussion and in many of the publications we have reviewed, is the notion that there is a problem in developing countries with the nutritional intakes of workers. This assumption has been challenged by some in the case of energy.

Energy deficiency is a strange beast which may be defined or measured in several ways (Figure 3). As Beaton (1985) points out, we can do this by focusing on either intake, stores, or expenditure. Different answers will be obtained about the gravity of the nutritional problem depending upon the selection of energy, stores, or expenditure as the aspect of choice and depending upon the specific criterion value used. Those clinically-oriented researchers have focused on energy stores as measured by weight for height and skinfolds. The observation that few adults are emaciated in developing countries except in famines has led some to propose that adults in less developing countries are "small but healthy" (Seckler, 1980).

Stores may be maintained, of course, at varying levels of intake and expenditure. Increasingly, nutritionists are beginning to focus on expenditure as perhaps a better measure of health (Arroyave, 1983; Beaton, 1983). Being sedentary helps neither the well-to-do in developed countries nor the unemployed in developing countries. As noted before, we have very little information but it seems energy supplements to workers, at least in Guatemala, cause them to be more active after work. These activities may be largely leisure but important nonetheless. Parent-child interaction, community involvement, and cottage industries are the sorts of things that might benefit. What some are saying is that in the final analyses, the specification of energy needs for individuals and populations must be a social definition (Arroyave, 1983; Beaton, 1985). As Beaton (1985) says, we must specify "energy requirements for what." We heartily agree that the "what" must include energy for maintenance, a full day's work schedule and socially acceptable discretionary activity. This line of reasoning will yield estimates of energy needs for populations which will be quite high and apt to be labelled as one more example of the persistent exaggeration of nutritionists (Poleman, 1981). What must be understood is that such "liberal" estimates are more goals for a healthy society than minimum levels for preventing starvation. Levels of estimated intakes to meet maintenance and minimal needs are also needed, but so are goals.

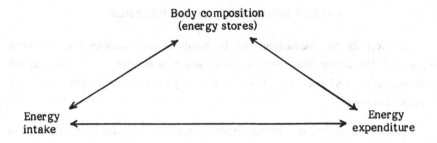

Figure 3:   The eternal triangle of energy
         (From Beaton, 1985.)

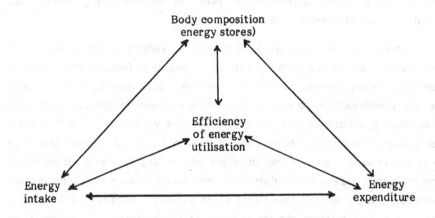

Figure 4:   Efficiency of energy utilisation: is it a significant factor in
               energy balance?

ENERGY DEFICIENCY

The views of Sukhatme and Margen (1982) have caused a great deal of discussion in the nutrition community.   Basically, they propose that there is a fourth factor in energy balance, namely energy efficiency (Figure 4).   They believe they have detected autoregulation in energy and protein intakes in individuals and propose that this is being achieved by adjustments in the metabolic efficiency of energy utilisation.   Sukhatme and Margen (1982) argue that through changes in metabolic efficiency individuals can vary energy expenditures without altering intakes and body stores.   That variability in efficiency exists is probably true but what is at issue is the extent of its importance. They propose that fixed levels of energy expenditure without changes in stores can be maintained over the range of $\pm$ 30 percent of the usually recognised requirement. If the mean requirement is 3000 kcal/day, that would mean a range from 2100 to 3900 kcal/day. They also propose that this feat is achieved without any loss in function.

The ideas of Sukhatme and Margen (1982) have been strongly criticised (Beaton, 1985; Arroyave, 1984; Rand & Scrimshaw, 1984). There is no evidence for such a broad range of adaptability. Also, a wealth of data from developing countries indicates that when energy expenditures rise and intakes do not meet the rise, that body stores do change. We see this, for example, in the case of seasonal variations in body weight in some developing countries. Data from the Gambia also indicate that mothers deliver very small babies and reduced amounts of milk in the season of poor food availability and high energy expenditure but have substantially better reproductive performance in better times (Prentice et al, 1981). These examples illustrate that there are limits to adaptations to low energy intakes and that there are clear losses in function. Therefore, we cannot take the views of Sukhatme and Margen (1982) without question.

CONCLUSION

In conclusion, we would like to focus on the notion that improved nutrition leads to greater productivity and earnings in workers. Figure 5 presents some of the ideas expressed by Alan Berg of the World Bank and we would like to comment on the extent to which we believe these relationships are true. The data we have reviewed strongly suggest that adequate nutrition, both in the developmental phase and in adulthood, results in stronger, more energetic workers. Malnutrition and mortality, particularly in infants and small children, are clearly linked and hence improved nutrition undoubtedly lengthens the working life span. Severe malnutrition - and perhaps moderate malnutrition as well - do decrease cognitive skills, but the effect this may have on work output remains unknown. Improved nutrition, as suggested by a number of the studies we reviewed, may lead to fewer days lost to illness. Improved nutrition in children seems to reduce not so much the incidence of infections but the duration of episodes and their severity. We know little about these relationships in adults.

Through the mechanisms outlined in Figure 5, all reasonable, but not all documented, improved nutrition should result in what Immink, Viteri and Helms

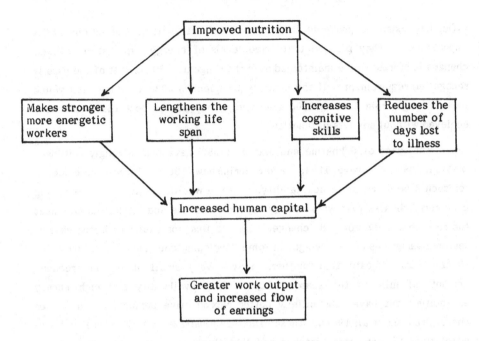

Figure 5: Nutrition, human capital, and earnings.

(1982) have called increased human capital. According to this view, malnutrition is a deterrent to economic development and interventions to improve nutritional status are justifiable on economic grounds.

But does increased human capital mean greater output? Hakim and Solimano (1976) believe this is not always the case. They write "The translation of improved individual capacities into greater individual and national productivity .... is largely dependent on the ability of the society to make effective use of such capacities. This in turn is a function of economic demands, social organisation, and the availability of complementary inputs. Enhanced individual capacities, whether they are brought about by better nutrition, education, or new technologies, may and often do go unused" (p. 250).

To us, the scientific evidence indicates that governments in developing countries must meet two challenges simultaneously. First, they must invest in programmes which will improve nutritional status and thus work capacity. Second, in order to realise the potential return from this investment in human capital, governments must also provide their people with the opportunity to be productive.

REFERENCES

Arroyave, G. (1983). Some issues relevant to the estimation of energy **needs** of humans. Draft paper prepared for the Berkeley/Stanford Colloquium on Nutrition and Rural Development, Berkeley, California.

Basta, S.S., Soekirman, M.S., Karyadi, D. & Scrimshaw, N.S. (1979). Iron deficiency anemia and the productivity of adult males in Indonesia. American Journal of Clinical Nutrition, **32**, 961-925.

Beaton, G.H. (1983). Energy in human nutrition: perspectives and problems. Nutrition Reviews, **41**, 325-340.

Beaton, G.H. (1985). The significance of adaptation in the definition of nutrient requirements and for nutrition policy. International Symposium on Nutritional Adaptation in Man, Royal Windsor, England, April 1-5, 1984, In: Waterlow & Baxter (eds.), Symposium volume, pp. 219-232. London: John Libbey.

Berg, A. (1981). Malnourished People: A Policy View. Policy and Basic Needs Series. Washington, D.C.: The World Bank.

Brooks, R.M., Latham, M.C. & Crompton, D.W.T. (1979). The relationship of nutrition and health to worker productivity in Kenya. East African Medical Journal, **56**, 413-421.

Brozek, J. & Schurch, B. (eds.) (1984). Malnutrition and Behaviour: Critical Assessment of Key Issues. Lausanne, Switzerland: Nestle Foundation.

Calloway, D.H. (1982). Functional consequences of malnutrition. Reviews of Infectious Diseases, **4**, 736-745.

Chen, L.C. & Scrimshaw, N.S. (eds.) (1983). Diarrhea and Malnutrition: Interactions, Mechanisms, and Interventions. New York/London: Plenum.

Davies, C.T.M. (1973). Relationship of maximum aerobic power output to productivity and absenteeism of East African sugar cane workers. British Journal of Industrial Medicine, **30**, 146-154.

Deolalikar, A.B. (1984). Are there Pecuniary Returns to Health in Agricultural Work? An Econometric Analysis of Agricultural Wages and Farm Productivity in Rural South India. Economics Program Progress Report 66. Andhra Pradesh, India: ICRISAT.

Gardner, G.W., Edgerton, V.R., Senewiratne, B., Barnard, R.J. & Ohira, Y. (1977). Physical work capacity and metabolic stress in subjects with iron deficiency anemia. American Journal of Clinical Nutrition, **30**, 910-917.

Heywood, P.F. (1974). Malnutrition and productivity in Jamaican sugar cane cutters. Ph.D. dissertation, Cornell University, Ithaca, New York.

Immink, M.D.C., Flores, R., Viteri, F.E., Torun, B. & Diaz, E. (1984). Functional consequences of marginal malnutrition among agricultural workers in Guatemala, Part II: Economics and human capital formation. Food and Nutrition Bulletin, **6**, 12-17.

Immink, M.D.C. & Viteri, F.E. (1981). Energy intake and productivity of Guatemalan sugarcane cutters: an empirical test of the efficiency wage hypothesis, Parts I and II. Journal of Development Economics, **9**, 251-287.

Immink, M.D.C., Viteri, F.E. & Helms, R.W. (1982). Energy intake over the life cycle and human capital formation in Guatemalan sugarcane cutters. Economic Development and Cultural Change, **30**, 351-372.

Johnston, F.E., Borden, M. & MacVean, R.B. (1973).   Height, weight and the growth velocities in Guatemalan private school children of high socio-economic class. Huamn Biology, **45**, 627-641.

Khosla, T. (1968).   Unfairness of certain events in the Olympic games.   British Medical Journal, **4**, 111-113.

Martorell, R. (1985).   Child growth retardation: a discussion of its causes and its relationship to health.   International Symposium on Nutritional Adaptation in Man, Royal Windsor England, April 1984, In: Waterlow & Baxter (eds.), Symposium volume, pp. 13-29. London: John Libbey.

Martorell, R. & Habicht, J.-P. (1986).   Growth in early childhood in developing countries, In: F. Falkner & J.M. Tanner (eds.), Human Growth: A Comprehensive Treatise, 2nd edn., pp. 241-262.   New York: Plenum.

Martorell, R., Delgado, H.L., Valverde, V. & Klein, R.E. (1981).   Maternal stature, fertility and infant mortality. Human Biology, **53**, 303-312.

Poleman, T.T. (1981).   A reappraisal of the extent of work hunger.   Food Policy, **6**, 236-252.

Pollitt, E. (1980).   Poverty and Malnutrition in Latin American Early Childhood Intervention Programs.   New York: Praeger.

Popkin, B.M. (1978).   Nutrition and labor productivity.   Social Science and Medicine, **13C**: 117-125.

Prentice, A.M., Whitehead, R.G., Roberts, S.B. & Paul, A.A. (1981).   Long-term energy balance in child-bearing Gambian women.   American Journal of Clinical Nutrition, **34**, 2790-2799.

Rand, W.M. & Scrimshaw, N.S. (1984).   Protein and energy requirements – insights from long-term studies.   Bulletin of the Nutrition Foundation of India, **5**, 1-2.

Rao, D.H. & Sastry, J.G. (1977).   Growth pattern of well-to-do Indian adolescents and young adults.   Indian Journal of Medical Research, **66**, 950-956.

Ricciuti, H.N. (1981).   Adverse environmental and nutritional influences on mental development: a perspective.   Journal of the American Dietetic Association, **79**, 115-120.

Satyanarayana, K., Nadamuni Naidu, A. & Narasinga Rao, B.S. (1980).   Adolescent growth spurt among rural Indian boys in relation to their nutritional status in early childhood. Annals of Human Biology, **7**, 359-365.

Satyanarayana, K., Nadamuni Naidu, A., Chatterjee, B. & Narasinga Rao, B.S. (1977).   Body size and work output.   American Journal of Clinical Nutrition, **30**, 322-325.

Satyanarayana, K., Rao, D.H., Rao, D.V. & Swaninathan, M.S. (1972). Nutritional working efficiency in coalminers.   Indian Journal of Medical Research, **60**, 1800.

Seckler, D. (1980).   "Malnutrition": an intellectual odyssey.   Western Journal of Agricultural Economics, **5**, 219-227.

Selowsky, M. & Taylor, L. (1973).   The economics of malnourished children: an example of disinvestment in human capital.   Economic Development and Cultural Change, **22**, 17.

Spurr, G.B. (1983).   Nutritional status and physical work capacity.   Yearbook of Physical Anthropology, **26**, 1-35.

Spurr, G.B., Barac-Nieto, M. & Maksud, M.G. (1977).   Productivity and maximal oxygen consumption in sugar cane cutters.   American Journal of Clinical Nutrition, **30**, 316-321.

Strauss, J. (1984). Does better nutrition raise farm productivity? Center Discussion Paper No. 457. New Haven, Connecticut: Yale University Economic Growth Center.

Sukhatme, P.V. & Margen, S. (1982). Autoregulatory homeostatic nature of energy balance. American Journal of Clinical Nutrition, **35**, 355-365.

Tanner, J.M., Hayashi, T., Preece, M.A. & Cameron, N. (1982). Increase in length of leg relative to trunk in Japanese children and adults from 1957 to 1977: comparisons with British and Japanese Americans. Annals of Human Biology, **9**, 411-423.

Viteri, F.E. (1974). Definition of the nutrition problem in the labor force, In: N.S. Scrimshaw & M. Behar (eds.), Nutrition and Agricultural Development. New York/London: Plenum.

Viteri, F.E. (1982). Nutrition and work performance, In: N.S. Scrimshaw & M.B. Wallerstein (eds.), Nutrition Policy Implementation: Issues and Experience. New York/London: Plenum.

Viteri, F.E. & Torun, B. (1974). Anemia and physical work capacity. Clinics in Haematology, **3**.

Wolgemuth, J.C., Latham, M.C., Hall, A., Chesher, A. & Crompton, D.W.T. (1982). Worker productivity and the nutritional status of Kenyan road construction laborers. American Journal of Clinical Nutrition, **36**, 68-78.

THE ROLE OF WORKING WOMEN IN A RURAL ENVIRONMENT WHEN NUTRITION IS MARGINALLY ADEQUATE: PROBLEMS OF ASSESSMENT

J.V.G.A. DURNIN and S. DRUMMOND

Institute of Physiology, University of Glasgow, Glasgow, U.K.

INTRODUCTION

This topic will be dealt with under the second of the two 'terms of reference' of this Symposium, which states that one of the objectives is "to assess critically methodology, to identify locations and research problems for field study, and to make proposals for future multidisciplinary investigations on work capacity in the tropics." This paper will, it is hoped, stimulate ideas and discussion about how nature and degree of the problems associated with marginally adequate nourishment can be investigated in women who have both to run a household and to undertake labour as well.

Much of this presentation is hypothetical - and a sad reflection that is on all the nutritional research which has been conducted on peoples in developing countries during the past 20 or 30 years. We have apparently only recently become aware, in any quantitative scientific manner, of the type of reactions that may occur in marginally nourished populations. In such situations over large parts of the world, in areas where food is not plentifully available, its supply in adequate quantities is spasmodic, dependent primarily on the season of the year - whether pre- or post-harvest. Moreover, it may also be very much affected by the bounty or otherwise of the harvest - as is all too obvious over much of north and central Africa.

In this context flagrant malnutrition is not being considered, but a marginal state where some adaptation might be expected to occur to these severe and recurrent nutritional stresses.

It is at least possible that women may be the most obvious members of the family to demonstrate the need for adaptation. When food is restricted, because the male head of the household may often be the principal wage-earner, he may properly receive preferential treatment.   His diet may not be affected to the same degree as the rest of the family.   On the other hand, the mother, who will also be faced with the problem of feeding her children, may well have to exist on limited rations.   If, as well, she has to undertake forms of work outside her duties of running the household, of preparing and cooking the food, and caring for the children - work such as toiling in the fields, tending a garden, looking after animals - she will be subjected to extra stresses and she may appear to require some form of adaptation for the situation to be tolerable in the long-term.

Adaptation is a very fashionable word.   In varying forms, adaptation is needed when the desirable energy requirement of the individual is not matched by the energy content of the diet.   The energy requirement is the level of energy intake from food that will balance energy expenditure when the individual has a body size and composition, and a level of physical activity, consistent with long-term good health.   The level of physical activity should include not only the economically and socially necessary quantity, but also any physical activity of leisure pursuits as well.   In the case of women, there is the added requirement of extra energy for pregnancy and lactation.

When all of these needs are inadequately met - which will usually be the case in situations of marginal nourishment for rural women such as have been described above - some forms of adaptation would be expected to occur in the individual.   This adaptation may conceivably mean a new equilibrium state with the inadequate food energy, but is more likely to lead to a state of slow but progressive deterioration.   For example, if food energy is insufficient, the individual woman may lose a little body mass, and then enter a more-or-less steady state at this lower level for the duration of the restricted food supply.   If the situation is more severe, the body mass may progressively decline until food once again becomes more plentiful.

Adaptations to insufficient food energy can be of several different kinds - biological or social and behavioral.   The biological adaptations in themselves can be of divergent forms.   In adults, they can consist simply of a decrease in body mass.   This has a double benefit; one consists of using the body reserves as a source of energy, and the other of reducing the energy requirement because of the smaller body mass (other things being equal, a woman weighing 45 kg will need less energy to remain in a steady state than a woman of 50 kg).

However, other biological adaptations may also happen.   The metabolic rate may lessen as a consequence of a smaller energy intake.   This has been well documented in states of semi-starvation, or even when food intake is reduced by about 500 kcal/day (2.1 MJ). Whether it occurs to any significant extent with smaller reductions in energy intake is not, to my mind, clearly proven yet.

Another biological adaptation - which may have a social/behavioral component - is a lessened amount of physical activity when food is restricted. This may affect only the optional physical activity of leisure, or it may impinge on economic activity as well, with the result that the woman is unable to carry out her various tasks in her normal fashion.

Social/behavioral adaptations could take various forms, some of which would be very difficult to measure. One adaptation might be that in the different circumstances where physical activity takes place, these could continue with superficially little alteration but the activity might be carried out at a slightly slower pace with a lower expenditure of energy. Alternatively, clear reductions could occur: e.g. the women might no longer go to work in the fields. Even this might be difficult to analyse since, at periods of the year when food might be limited, there may be less work to be done in the fields.

Similarly, alterations in leisure activity may happen, but these could equally be masked by perhaps long-standing cultural adaptations with changing lifestyles at different periods of the year, so that social physical activity could parallel food availability.

The energy expenditure of rural women may well be relatively high. There is plentiful evidence that women may have to expend considerable quantities of energy in agricultural tasks, and in household duties. These can amount to between three and four times the basal metabolic rate and in a normal working day can therefore increase energy needs by large amounts.

An extra biological stress for most women in these situations is to have several pregnancies and indeed they may spend a large part of their reproductive life being either pregnant or lactating. Pregnancy and lactation are nutritional areas where much confused and emotional thought exists, complicated by poorly controlled scientific investigation. Usually the extra requirements for these states are calculated on the basis that the extra energy theoretically needed to manufacture and maintain the fetus, enlarge the uterus, form the placenta, lay down some extra adipose tissue, increase the blood volume, and so on, must be equal to the actual extra energy needed by the mother. Little or no allowance is made for any physiological adjustment during pregnancy and lactation, which would reduce this theoretical requirement in any way, and almost no regard is paid to the considerable mass of data showing that something different actually occurs, because healthy pregnant women do not normally eat food containing an **extra** 200-300 kcal per day (0.8-1.3 MJ) during pregnancy, nor do women usually take in an extra 500 kcal (2.1 MJ) during lactation, which would be the theoretical requirement.

This problem has recently been studied and as part of a coordinated project in two industrialised countries (Scotland and Holland) and three developing countries (Thailand, the Gambia, and the Philippines) there are to date longitudinal data on more than 160 women in Scotland and Holland and more than 100 in the developing countries.   These women have had various measurements done on them - food intake, body mass and composition, levels of physical activity, BMR, metabolic rate during standardised exercise - all of these done serially, either two-weekly, four-weekly, or six-weekly, right throughout pregnancy beginning in the very early stages.   It is quite clear that several adaptations have taken place and the extra energy needed for a healthy mother to produce and feed a normal healthy baby is nowhere near the theoretical requirements.

This example has some relevance to women living in a rural environment where food supply is occasionally inadequate, for it illustrates the great need to study problems such as these with care and proper scientific control.   In this context, there is a research project which has been functioning in three developing countries, and which illustrates the methodological and other problems which need to be solved before acceptable data can be obtained.

The project (financed by a grant from the EEC) is taking place in India, in Begin with the collaboration of Professor Hautvast from Holland, and in Ethiopia with the collaboration of Professor Ferro-Luzzi from Italy.   One of the aims of this study was to examine some of the different adaptations which might occur in a group of women, living in a rural environment in South India, where it is already known that food availability will vary throughout the year with occasional periods when it will be less than adequate.   Since most of the study is a very intensive one, the number of women who can be investigated is limited to a relatively small number, i.e. about 50 women. A larger number, about 100, have had some simple anthropometric measurements taken serially throughout the year. They have been selected from a fairly large village population and it is hoped are representative of them, even though not randomly selected.   The general aim of the study has been to take many different measurements of these women throughout slightly more than one complete year.

In relation to the precise type of woman selected, it was decided to use adult women in the childbearing period with a maximum 'normal' work load resulting from domestic and other duties. They were either married or widowed, with children to care for.   Because it would have complicated the analysis to study women at varying stages of pregnancy and lactation, concentration was on non-pregnant, non-lactating women.

As part of the necessary characteristics of the sample, the women had some involvement in agricultural tasks, either as seasonal labour, work in their own gardens, or looking after animals. They were reasonably representative of the average socioeconomic status of the village, and included mostly the poorer women with a small number of 'middle-class'. They were in reasonable health without clinical signs of tuberculosis, diabetes, liver disease, etc., none being physically handicapped.

The measurements made related to variables such as alterations in body weight and body composition throughout the year, alterations in food intake, changes in the duration and type of physical activity, in basal metabolic rate, and variations in their physiological capacity for physical activity. Some of these measurements involved considerable practical problems.

Food intake was measured using the individual inventory technique, which means that all of the food eaten by each individual woman was weighed and recorded immediately before being consumed, together with a complete description of the food; the energy and nutrient content of the diet were calculated using tables of the composition of foods.

Whilst it was intended to repeat this study on each individual at intervals of about six weeks, delays in obtaining Indian Government approval for the project meant that in fact the study was done on only two, or occasionally three, times in all in the attempt to cover seasonal variation.

The field workers consisted of one expatriate postgraduate student, who spent most of each day in the village, assisted by 6-8 young literate girl helpers who came from the immediate environment, and a locally-trained qualified dietitian. The field workers and the dietitian were provided through the good offices of the Director and of Dr. Satyanarayana, of the National Institute of Nutrition in Hyderabad. The field work was initially organised by the Institute. These arrangements should allow for six women to be studied each week, permitting the total sample of about 50 women to be covered during approxmately two and a half months. The dietary measurements use battery-operated digital dietary scales of 2 kg capacity and 5 g precision calibrated regularly. (These scales have a zeroing button and are relatively easy to use, in practice, with a high degree of accuracy.)

Since changes in body weight and body composition were critical to the assessments, a fairly complete anthropometric set of measurements was done at the beginning of the study. These included height, weight, skinfold thicknesses, limb circumferences, and bone diameters. Body weight was measured on 100 kg capacity battery-operated digital scales. Weights were taken daily during each dietary survey and also at the time of the measurements of basal metabolic rate (done approximately half-way between the food intake assessments). Skinfold thickness and limb circumference will be measured once during each dietary survey.

A larger sample of women was also included where the only measurements, repeated three or four times throughout the year, would be of body weight and skinfold thicknesses. This allowed the assessment not only of how representative the intensively-studied sample was of the larger village population, but also gave information on weight and body fat changes.

Since alterations in physical activity are obviously important there was a detailed 'time and motion' study of the activities of the women, and a qualitative assessment of the energy cost (e.g. walking fast or slow, working slowly or at moderate rates, etc.). The socio-economic cultural value of the activity could clearly have relevance, for example washing for the family, washing for other people in the community for payment, cooking for the family alone, or for guests, and so on. These determinations, although borne in mind, were too difficult in practice, and should have required the professional assistance of a social anthropologist. The time and motion records were gathered by an observer (not by the woman herself) and the young local assistants were involved both in this and in the diet survey.

Basal metabolic rate determinations may also be important, since they provided data on whether or not a biological adaptation may be taking place in response to restricted food intake. The measurements were obtained using the Douglas bag and oxygen consumption technique, and were repeated, with the appropriate methodological care, twice throughout the year with the measurements taking place two or three weeks apart from the dietary survey.

The energy cost of the important physical activities were measured using a Douglas bag. These activities were planned to include sitting, standing, moving around doing housework, walking at different rates, and any significant work activity. Again, in practice, it is very difficult to obtain the quantitative data which are really required in this type of study. There is no ideal field technique for measuring energy expenditure and indeed it is very difficult to be sure that even several measurements of energy expenditure really reflect the normal situation for the individual. Many assumptions have to be made using this type of methodology, and there are considerable problems involved in analysing properly the resultant data.

Finally, a simple step test was carried out to measure exercise capacity, which involved submaximal levels of exercise, at three different levels. These three levels required (1) a heart rate of about 100 beats per minute, (2) one where the heart rate was about 130 beats per minute, and (3) a level with a heart rate of about 160 beats per minute. Using these three sets of measurements, with oxygen consumption being simultaneously determined, it proved possible by regression analysis to predict the maximal exercise capacity of the woman. A step test of this nature has been validated in Glasgow, when comparison with a similar exercise test on the treadmill was made. In the present study, measurements have been made of about 30 women and showed fairly good agreement between the two sets of assessments. It is hoped that a step test carried out in this manner will provide information of almost equal value to the equivalent test done on a treadmill in a well-equipped laboratory.

Various other pieces of information of sociocultural and biological interest need to be collected in a study such as has been described. These include data on the menstrual cycle, pathological events (illnesses, accidents, etc.), cultural events such as fastings or celebrations, the success or otherwise of the harvest, and climatic occurrences.

It is apparent that studies such as these are necessary to allow us to assess situations of marginal nutrition much more rigorously, but they are certainly difficult to carry out in practice with a high enough degree of accuracy, and we approach such studies with proper humility.

ACKNOWLEDGMENTS

This study was financed by grants from the European Economic Community and from the U.K. Overseas Development Administration.

DEFINING ANAEMIA AND ITS EFFECT ON PHYSICAL WORK CAPACITY AT HIGH ALTITUDES IN THE BOLIVIAN ANDES

J. D. HAAS[1], D. A. TUFTS[1], J. L. BEARD[2]
R. C. ROACH[1] and H. SPIELVOGEL[3]

[1]Division of Nutritional Sciences, Cornell University, Ithaca, N.Y., U.S.A.
[2]Department of Nutrition, The Pennsylvania State University,
University Park, Pennsylvania, U.S.A.,
[3]Instituto Boliviano de Biologia de Altura, La Paz, Bolivia

INTRODUCTION

Of all the ecozones in the tropics, one of the most challenging in terms of its effects on physical work capacity is found in the high altitude mountains. Not only does one have to contend with many of the health and nutritional problems associated with underdevelopment but one also has to adjust to the reduced atmospheric pressure and partial pressure of oxygen associated with high elevations.   Since oxygen is required for all aerobic work, a reduction in oxygen tension will reduce aerobic work capacity unless substantial adaptations occur in systemic oxygen transport and utilisation.

Considering that all of the tropical high altitude populations also reside in less developed and economically poor countries, it is very likely that many environmental stressors act in concert to reduce the physical performance of these indigenous populations.   Most previous studies of work capacity in high altitude natives of the South American Andes suggest that the maximum aerobic capacities are comparable to indigenous low altitude groups engaged in similar types of physical activities (Buskirk, 1978).   This suggests that Andean natives have adapted to the stresses of hypobaric hypoxia to a large extent through improved oxygen transport and utilisation.   This may be true for the samples studied.   However, if one examines the physical characteristics of weight and height for these highland samples, there is some question as to how repre- sentative they are of the general population of highland natives.   Figure 1 presents mean weights plotted against mean heights for various samples of indigenous Amerindians living above 3500 meters in the Central Andes.   In all

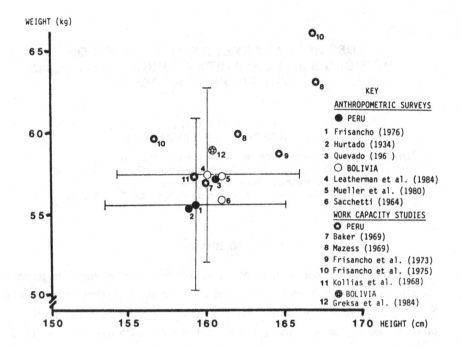

Figure 1:   Mean weights and heights of various samples of native Andean
              males. Intervals represent ± 1 SD for the larger and more
              representative surveys, numbers 1 and 4.

but two studies (Baker, 1969; Kollias et al, 1968), the samples from which we
have work capacity data are taller or heavier than men of similar age from
several anthropometric surveys of the general Andean native population. If
greater weight and stature are indicative of better health and nutrition conditions and increased European genetic admixture, then the exercise studies
appear to represent the better-off segments of the highland population.   Considering the documented effects of poor health and various types of malnutrition
on work capacity (Spurr, 1984), it is very likely that the general Andean
population has lower work capacities than have been suggested by the current
literature.   To our knowledge, no research has been conducted at high altitudes
on the relationship between malnutrition and physical performance.

        Most of the research at low altitudes on the effects of malnutrition and
work capacity has focused on protein energy malnutrition (PEM) and nutritional
anaemias.   Both classes of nutrition problems are of potential importance to
tropical high altitude populations.   While PEM in adults at high altitude is

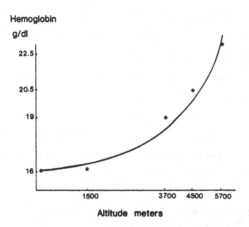

Figure 2:   Relationship between haemoglobin concentration and altitude in Peruvian males. (Adapted from Hurtado et al, 1945.)

probably assessed in similar ways as at low altitude and has similar consequences on work capacity as has been shown at low altitude, there is reason to believe that this is not the case for nutritional anaemias.   It is well known that haemoglobin concentrations are elevated at altitudes above 3000 meters in the South American Andes (Garruto, 1976; Garruto & Dutt, 1983).   This results from increased haemopoiesis as one of several physiologic responses to hypobaric hypoxia.   Figure 2 demonstrates the altitude effects on mean haemoglobin values for adult men in South America.   This adaptive response is most noticeable in newcomers to high altitudes and may be less extreme but still present in Andean natives (Garruto & Dutt, 1983) and perhaps non-existent in Himalayan populations (Beall et al, 1983).   Since iron deficiency is known to limit haemopoiesis at low altitudes, it may be considered a serious impediment to successful adaptation at high altitudes.   However, relatively little research has been conducted on the prevalence of iron deficiency at high altitudes. Moreover, our knowledge at high altitudes of the effects of iron deficiency on haemopoiesis, as well as the criteria for anaemia and the functional conse-quences of iron deficiency and anaemia are not clearly understood.   Since 1980 we have been examining many of these issues in a high altitude population residing between 3500 and 3800 meters in La Paz, Bolivia.   Our research strategy has been first, to establish criteria for describing the normal levels and distribution of haemoglobin;   second, to determine the prevalence of iron-deficiency in the population;   third, to determine the relationship between iron-deficiency and anaemia, and fourth to determine the functional consequences of anaemia.   This paper presents some of the results of this ongoing research, summarising information on the first three steps and providing more detail on our findings regarding physical work capacity.

Before presenting the research results, it may be helpful to review briefly what others have found regarding normal values of haemoglobin and prevalence of iron deficiency and anaemia at high and low altitudes. Also, a brief review of research on the relationship between iron deficiency, anaemia and work capacity at low altitudes will be presented, since much of this work serves as a rationale for our research at high altitude in Bolivia.

DEFINING NORMAL HAEMOGLOBIN LEVELS AND PREVALENCE OF ANAEMIA AND IRON DEFICIENCY

Prior to examining the functional consequences of abnormally low haemoglobin concentrations, it is necessary to define the normal and abnormal populations with respect to haemoglobin. There are generally two approaches to defining normal: one either attempts to define a normal population, or a normal individual.   Often, the two approaches are combined in that a specific cut-off value for haemoglobin is defined from population statistics (below 5th percentile, below minus 2 standard deviations, etc.) so that all individuals below that cut-off value are considered anaemic.   It is important to recognise, however, that anaemia is defined functionally in terms of reduced oxygen carrying capacity. Therefore, individuals with haemoglobins above the cut-off may be functionally anaemic but not classified as such, and individuals below the cut-off may be functionally normal but classified as anaemic.   The development of cut-off values for medical and nutritional diagnosis has been widely discussed (Galan & Gambino, 1975; Habicht et al, 1982) and is not suitable for elaboration in this paper.   The conventional approach to decribe anaemia for a population is to estimate a prevalence of those individuals below a specific cut-off value.   The World Health Organization (1968) has suggested that values below 13 g/dl for adult men, 12 g/dl for non-pregnant women, 11 g/dl for pregnant women and 12 g/dl for primary school age children, be diagnostic of anaemia.   However, considering the positive altitude effects on mean haemoglobin (Figure 2) several authors have attempted to adjust the cut-off values for diagnosing anaemia at high altitudes.   Some examples of these altitude- pecific cut-off values are presented in Table 1 for primary school children at 3700 metres.   It is clear that good agreement on cut-off values does not exist.   These cut-'ff values were applied to a sample of 641 children in La Paz (Quinn, 1982).   Depending on the cut-off value one accepts, the prevalence of anaemia varies from zero to 23%.

Table 1:   Estimates of anaemia prevalence using several recommended high altitude haemoglobin criteria

Source	Haemoglobin cut-off criteria	Number of children below cut-off	% Anaemic in total group (N=641)
Sauberlich et al, 1974	<11.3 g/dl	0	0.0%
ICNND, 1964	<12.3 g/dl	2	0.3%
Dallman et al, 1980	<13.8 g/dl	12	1.9%
US AID, 1979	<15.0 g/dl	150	23.4%

Several methods were employed in our research to identify the normal range and distribution of haemoglobin and to confirm our definition of normality. The first method involved the plotting of cumulative frequency distributions of haemoglobin for selected subgroups of the population defined by sex, age and physiological status.   Haemoglobin in healthy well-nourished populations fits a Gaussian or normal distribution (Garby et al, 1969; Cook et al, 1971; Meyers et al, 1983).   If the cumulative frequency of haemoglobin values is plotted on probability paper it produces a straight line with the mean equal to the median or 50th percentile and plus or minus one standard deviation represented by the 84th and 16th percentiles, respectively.   However, many sampled populations contain a subgroup of individuals with low haemoglobins associated with iron deficiency anaemia.   This subpopulation has its own distribution of haemoglobin which may or may not be Gaussian.   When two subpopulations are combined, the distribution becomes skewed and the probability plot deviates from linearity at the low end of the range.   The size of the anaemic group can be estimated by the degree of deviation that is seen in the tail of the distribution.

This approach was applied to haemoglobin data from eight Latin American countries by Cook and colleagues (1971) to determine the prevalence of anaemia and the mean and standard deviation of the non-anaemic population.   Their results are summarised in Figure 3 for pregnant and non-pregnant women and adult men.   All three distributions have 'tails' which reflect an admixed anaemic subpopulation.   The estimated size of the anaemic subpopulations within the total populations of pregnant women, non-pregnant women and men were 22%, 12% and 2%, respectively.   The means shown in the figure are for the non-anaemic subpopulations, as reported by Cook et al (1971).

We have applied a modified version of this approach to estimate normal parameters and prevalence of anaemia in four population groups, all

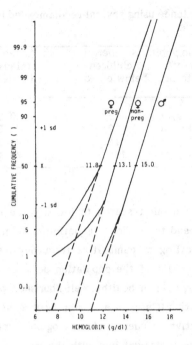

Figure 3:   Distribution of haemoglobin in three adult low altitude populations
from Latin America.   (Redrawn from Cook et al, 1971.)

permanently residing at a mean altitude of 3700 meters in La Paz. The groups
are 529 adult male labourers, 400 adult non-pregnant women, 130 pregnant
women examined during the eighth post-menstrual month, and 641 primary
schoolchildren between six and ten years of age.   Nearly all subjects were born
at high altitude and none had made a recent sojourn to lower altitudes prior to
examination.   All groups were represented by indigenous Amerindians, as well
as individuals of European and mixed European-Indian ancestry.

Data from the sample of adult male labourers are presented in Figure 4.
The cumulative frequency distribution indicates that deviations from linearity
exist at both extremes of the distribution thus suggesting the presence of three
subpopulations.   These are interpreted 'to represent anaemic, normal and
polycythaemic groups, the latter being associated with pathologies of oxygen
transport that lead to an over-production of red blood cells and excessively high
haemoglobin concentrations.   Estimates of the prevalence of anaemia and
polycythaemia based on the areas between the extrapolated straight line of the
normal middle population and the empirical tails gives values of 0.8% and 4%,
respectively (Tufts et al, 1985).

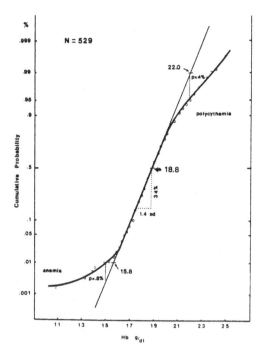

Figure 4:   Distribution of haemoglobin for adult men from La Paz, Bolivia, at
              3600m altitude.   18.8 g/dl indicates the mean with a SD of 1.4 for
              the population of normal, non-anaemic and non-polycythaemic men.
              ˙(From Tufts et al, 1985.)

Replotting the data for this same population after excluding all iron-
deficient individuals with transferrin saturation (TS) less than 15% results in
changes in the distribution as shown in Figure 5.   After removing the iron-
deficients, the lower tail of the distribution is lost, confirming that most anemia
is due to iron deficiency in this subpopulation.   Note that the tail remains
unchanged at the upper range of the haemoglobin distribution.

The same analysis was applied to all four subgroups from La Paz.   The
results are summarised for the four high altitude groups from La Paz and the low
altitude populations from elsewhere in Latin America in Table 2.   The important
points to note regarding haematologic variation are: (1) after adjusting the
original means by removing anaemics and polycythaemics, the high altitude
values remain systematically higher than low altitude values; (2) the prevalence
of iron deficiency is somewhat variable between age and sex groups, with women
and especially pregnant women being more iron-deficient than men; (3) considering the relative pattern of prevalence of iron-deficiency at high altitude, there is

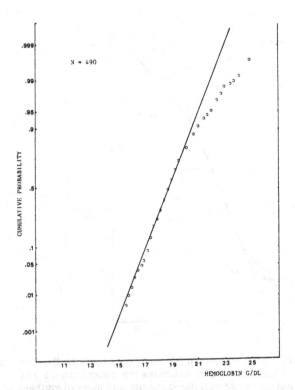

Figure 5:   Distribution of haemoglobin for adult men with transferrin saturation above 16%.   This is the same population as in Figure 4, after
removal of 39 cases diagnosed as iron-deficient, TS < 15%.
(From Tufts et al, 1985.)

a surprisingly low prevalence of anaemia, especially among pregnant women; and
(4) polycythaemia is only seen at high altitude and then only in adults.

It should be noted that the approach used by Cook et al (1971) for the low
altitude samples and by us for the high altitude samples are not identical, thus
the results are not directly comparable.   The relatively higher prevalence of
iron deficiency and anaemia in the low altitude samples of women may result in
an under-estimation of the mean and standard deviation of the normal population
which results when the large abnormal portion is removed (Meyers et al, 1983).
Thus the altitude differences in mean haemoglobin may be less than indicated in
Table 2.   Also, the estimated prevalence of anaemia may be similarly affected.
A large subpopulation of true anaemics will tend to reduce the slope of the total
population frequency distribution line in Figure 3, leading to an under-estimation
of prevalence based on the deviation from the straight line.

Table 2: Haemoglobin, polycythaemia, anaemia and iron deficiency in high and low altitude populations.

	N	Haemoglobin, g/dl		Prevalences, %		
		unadjusted mean	adjusted[1] mean ± SD	Poly-cythaemic[2]	Anaemic[2]	Iron deficient[3]
Low altitude (Cook et al, 1971)						
Men	304	15.0	15.1 ± 1.1	0	2	3
Women – nonpregnant	485	13.1	13.2 ± 1.0	0	12	21
Women – pregnant	899	11.4	11.8 ± 1.2	0	22	49
High altitude (3700m)						
Men	600	19.0	18.8 ± 1.4	4	1	9
Women – nonpregnant	400	19.0	18.8 ± 1.4	2	2	12
Women – pregnant	130	13.9	13.9 ± 1.4	1	4	41
Schoolchildren	641	15.6	15.6 ± 0.8	0	1	9

[1] From central Gaussian distribution, excluding anaemic and polycythaemic subpopulations.
[2] Deviation from Gaussian distribution.
[3] Transferrin saturation <16%.

Prevalence estimates of iron deficiency are based on transferrin saturation below 15%. While the criteria was applied for all groups studied, there is a possibility of considerable interlaboratory variation in the techniques employed to estimate serum iron and total iron binding capacity. We have no way of knowing whether systematic technical error accounts for the observed altitude differences in prevalence of iron-deficiency. Theoretically, there appears to be no reason to expect different cut-off values for transferrin saturation between high and low altitudes.

The apparent discordance between high prevalences of iron deficiency and low prevalence of anaemia may be due to our exclusive use of transferrin saturation as an indicator of iron status. A transferrin saturation less than 15 percent may lead to misdiagnosis of chronic iron-deficiency. Acute inflammatory disease or reticuloendothelial blockage may limit iron delivery to the bone marrow and reduce transferrin saturation temporarily, but not result in a functional impairment in haemoglobin concentration. Chronic inflammation and reticuloendothelial blockage could lead to anaemia by reducing haemopoiesis even with adequate iron in storage. The use of other indicators of iron status, such as serum ferritin and erythrocyte protoporphyrin along with transferrin saturation would greatly improve the diagnosis of iron deficiency (Cook & Finch, 1979).

The relatively low prevalence of anaemia among pregnant women at high altitude warrants further investigation. Indirect evidence from cross-sectional data in Bolivia suggests differential effects of altitude on plasma volume expansion during late pregnancy. There appears to be a greater plasma volume expansion at high altitude, whereas the prevalence of anaemia and iron deficiency is greater at low altitudes.

Finally, the public health significance of polycythaemia needs to be studied more. Our data indicate an increased risk of polycythaemia related to increasing age and overweight. There may also be ethnic differences in the prevalence of polycythaemia (Garruto & Dutt, 1983). We have also investigated the functional consequences of high haemoglobin concentrations (above 20 g/dl in terms of work capacity and oxygen transport. The results of this analysis are presented below with the analysis of anaemia.

The prevalences of anaemia in Table 2 were determined by the mixed distribution analysis and not by calculating the percentage of individuals below a cut-off value. However, for health screening of individuals and simple population estimation it is often useful to employ cut-off values. There is no simple method for ascribing a certain cut-off value to the haemoglobin distribution; the approach depends on the purpose or use of the cut-off. With this in mind, we did attempt to define several cut-off values which were based on probability of deviation from a normal curve. This exercise gave a cut-off value of 15.8 g/dl for diagnosing anaemia in adult men (Tufts et al, 1985). The definition of polycythaemia is somewhat problematic, since we have no test of causation comparable to iron-deficiency at the anaemic end of the distribution. Our examination of work capacity in this group used 20 g/dl haemoglobin concentration which is lower than the 22 g/dl indicated by analysis of the haemoglobin distribution. This choice was based primarily on consideration of sample size. Only one of our exercise subjects had haemoglobin values above 22 g/dl; such high values are usually associated with obesity, advanced age or pulmonary disorders that prevent administration of a demanding exercise test protocol. A validation of the cut-off values was carried out in a subsample of fifty-six men to determine whether anaemia and high haemoglobin as defined were related to reduced physical working capacity.

ANAEMIA, IRON DEFICIENCY AND WORK CAPACITY

We chose to evaluate the significance of haemoglobin variation in terms of its effect on one of several potentially important functional domains, physical working capacity. This functional domain is a very important one at high altitude because not only is it known to be reduced by hypobaric hypoxia, but it determines an individual's ability to sustain heavy manual labour and engage in economically important farming activities. Several studies have shown reduced physical work capacity in anaemic versus non-anaemic subjects (Gardner et al, 1977; Davies et al, 1973; Basta et al, 1979). The results of these studies suggest that if one aspect of the oxygen transport system, such as haemoglobin, is operating at suboptimal levels and there is limited physiological compensation from other components of the system, the work capacity should be reduced. Other studies reveal that tissue iron-deficiency, which usually accompanies anaemia, may also limit work capacity by reducing the functioning of oxidative enzymes needed to support muscular work (Finch et al, 1976, 1979; Dallman, 1982). Dallman (1982) suggests that iron-deficiency effects on tissue enzymes relate to reduced phyical endurance, while its effect on haemopoiesis relates to reduced aerobic work capacity at maximal exertion. We employed various measures of work capacity, including maximal aerobic power or VO_2 max. This is a measure of the overall ability of the oxygen transport system to deliver oxygen to working skeletal muscles while they are engaged in maximal exertion.

We also used the oxygen consumption at a cardiac frequency of 150 bpm (PWC_{150}). This computed value provides similar results and interpretation as the VO_2 max as long as age variation in the sample is not great. It has the advantage of providing valid data for individuals who may not have reached maximum exertion at the highest level of achieved physical work in the particular test employed.

Subjects were first tested with a progressive exercise protocol adapted from Balke and Ware (1959) on an electronically braked bicycle ergometer. The procedure called for three to five work loads each of two minutes duration until the subject reached maximum voluntary exertion and could not continue the test. The results of this test are presented for anemics and non-anemics in Table 3. Age, height and weight did not differ between groups. The PWC 150 data for these subjects are shown graphically in Figure 6. The VO_2 max was estimated for each subject by extrapolating the regression line of his individual VO_2 versus heart rate, obtained at the various submaximal work loads, to a maximum heart rate of 220 minus age (McArdle et al, 1981). The maximum aerobic power for anaemics and non-anaemics correcting for body weight is 35.3 ml/kg/min and 43.4 ml/kg/min, respectively. Anaemics have significantly lower work capacities than non-anaemics in all three expressions of the data. There were no observed differences in aerobic work capacity between subjects with haemoglobins above 20 g/dl and those with haemoglobins between 15.8 and 20 g/dl. The non-anaemics in this analysis include all subjects with haemoglobins above 15.8 g/dl.

A second exercise test was administered to determine various physiological responses in anaemics compared to non-anaemics at the same relative work loads. The protocol called for a series of steady state work loads defined as approximately fifty, seventy and eighty percent of work relative to each subject's VO_2 max as estimated from the first test. Work at each exercise level lasted five to seven minutes. A broad range of ventilatory, circulatory and metabolic parameters were measured during the steady state test. Only a few are presented here to describe the nature of the adaptations made by iron deficient anaemics as they exercise under hypobaric hypoxic conditions. The analysis was also extended to examine the exercise response of individuals with high haemoglobin concentrations, who could be marginally polycythaemic by virtue of having a haemoglobin over 20 g/dl. Thus, functional consequences of pathology associated with the extremes of haemoglobin distribution could be evaluated in three groups: anaemics, normals, and polycythaemics. The groups were further defined in that all anaemics were iron-deficient and all normal and

Table 3: Aerobic capacity of anaemics and non-anaemics as estimated by
individual regressions of HR and VO_2 (mean ± SD).
(From Tufts et al, 1985.)

	Anaemics n=10	Non-Anaemics n=46
PWC150 (VO_2, l/min)	1.54 ± .27	1.86 ± .34*
Estimated VO_2 max (l/min)	2.12 ± .36	2.69 ± .51*

* p <.05 - anaemics compared to non-anaemics (t-test).

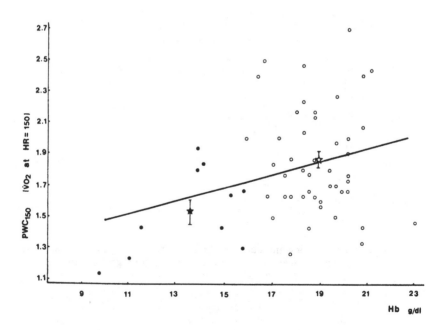

Figure 6: Relationship between oxygen consumption at a heart rate of
150 bpm and haemoglobin concentration in 56 adult male laborers
from La Paz. Filled circles indicate iron-deficient anaemics, with
haemoglobin <15.8 g/dl, transferrin saturation <15%, and serum
ferriten <12 ng/dl. Stars indicate means (±SD in brackets) for
anaemic and non-anaemic groups. Regression line fits the entire
data set. (From Tufts et al, 1985.)

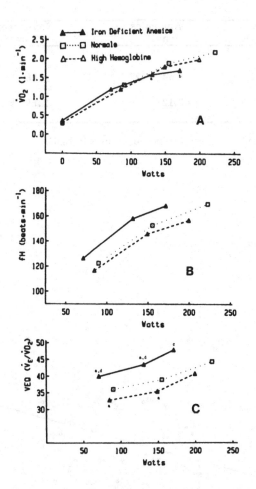

Figure 7:   Ventilatory and cardiac responses to steady state submaximal
exercise (watts) in anaemic, normal and high-haemoglobin male
subjects at high altitude (3600m).

    A: Oxygen consumption (VO_2);
    B: Cardiac frequency (fH);
    C: Ventilation equivalent (VEQ).

a = significantly different from normals, p <.05, ANOVA;
b = significantly different from normals, p <.01, ANOVA;
c = significantly different from high-haemoglobin, p <.05, ANOVA;
d = significantly different from high-haemoglobin, p <.01, ANOVA.

(From Roach, 1987.)

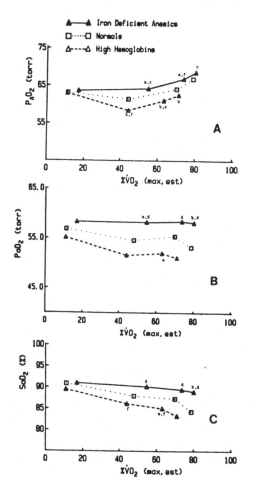

Figure 8: Oxygen tensions and saturations during submaximal exercise relative to percent of estimated maximum oxygen consumption (% VO_2 max est.) in anaemic, normal and high-haemoglobin male subjects at high altitude (3600m).

A: Alveolar oxygen (P_aO_2);
B: Arterial oxygen (P_aO_2);
C: Percent arterial saturation (S_aO_2).

a = significantly different from normals, p<.05, ANOVA;
b = significantly different from normals, p<.01, ANOVA;
c = significantly different from high haemoglobin, p<.05, ANOVA;
d = significantly different from high haemoglobin, p<.01, ANOVA;
e = significantly different from normals, p<.05, ANOVA;
f = significantly different from normals, p<.01, ANOVA.

(from Roach, 1987.)

high haemoglobin subjects are not iron–deficient.   There are no significant differences among groups in weight, height, age, weight/height and body surface area.

Figure 7A presents the VO_2 at different work loads for each group.   The symbols reflect data points equal to approximately 15, 50, 70 and 80% of VO_2 max from left to right along each line.   Anaemics have significantly lower oxygen consumptions at 70 and 80% of VO_2 max when compared to normal subjects, and completed the test at a lower work load than non–anaemics.   Also, at any work load anaemics have higher heart rates than the two non–anaemic groups, as can be seen in Figure 7B.

In order to determine the nature of the physiologic adaptations to work among the three groups, selected characteristics of oxygen transport and utilisation were examined.   Ventilatory adaptations are significant as shown in Figure 7C for ventilatory equivalent (VEQ).   Both low and high haemoglobin groups differ significantly from the normal subjects in the ventilation required per unit of oxygen consumption, with the normal group being intermediate.

The relative hyperventilation of the anaemic subjects is effective in raising alveolar O_2 tensions (P_AO_2) as shown in Figure 8A.   This same trend can be seen in Figure 8B for arterial oxygen tension (P_aO_2), with anaemics having significantly higher P_aO_2 than normal men at 50 and 80% of VO_2 max.   Arterial O_2 saturations, shown in Figure 8C, (S_aO_2) are also higher in anaemics than in normals but the difference is only significant at the higher work load (80% of VO_2 max).   Whether this is due to the inability of anaemics to desaturate their haemoglobin during heavy exercise, or due to enhanced oxygen affinity cannot be determined from these data.

Lung diffusing capacity for carbon monoxide (D_Lco) was determined at rest and during moderate exercise (80 watts).   The results are shown in Table 4. While no group differences were observed at rest, there is a tendency during exercise for lower D_L in anaemics and a higher D_L among those with high haemoglobins compared to controls.   It appears that most of this difference is due to variation in oxygen transport, since correction for haemoglobin concentration using the formulas of Dinakara et al (1970) results in anaemics having higher D_L relative to the other groups.   Studies of the effects of haemoglobin concentration on D_L through alternations in pulmonary vascular perfusion and O_2-haemoglobin association have not been done at high altitude, and may reveal possible mechanisms for adaptation to the double hypoxic stresses of anaemia and low atmosphere oxygen tension.

Table 4: Mean values (± S.D.) and ranges for $D_L$co and $D_L$co corrected* for haemoglobin in iron-deficient anaemic (IDA), normal (NORM), and high-haemoglobin (HHB) groups.

$D_L$co (ml/min/mmHg)	Group (n)		
	IDA (8)	NORM (21)	HHB (8)
Rest (observed)	23.1 ± 12.1 9—47	22.8 ± 10.9 9—48	21.6 ± 8.8 10—35
Rest (corrected)	26.6 ± 12.8[a,c] 9—48	18.0 ± 8.5 7—34	15.5 ± 6.2 7—24
Exercise (observed)	38.3 ± 10.2[d] 30—63	42.7 ± 7.7 29—62	48.9 ± 7.8 35—65
Exercise (corrected)	44.6 ± 11.3[b,d] 28—65	33.5 ± 5.8 22—45	33.8 ± 5.8 25—46

* corrected for haemoglobin concentration according to Dinkara et al (1970).

[a-d] ANOVA. a = significantly different from NORM, $p < .05$:
b = significantly different from NORM, $p < .01$;
c = significantly different from HHB, $p < .05$;
d = significantly different from HHB, $p < .01$.

Cardiac output at 50 and 70% VO_2 max is not different between groups, although arterial O_2 content (C_aO_2) and systemic oxygen transport (SOT) are significantly lower in anaemics compared to normals. Thus, it appears that whatever adaptations the anaemic subjects have made to exercise, they are still not able to fully compensate for the lower oxygen carrying capacity of the blood.

Some metabolic responses also distinguish the groups. Blood arterial pH drops equally in all groups during exercise, as does bicarbonate, and no intergroup differences were observed. Lactate increases similarly in all groups when compared at the same level of work relative to VO_2 max (Figure 9A). However, assuming no differences in total blood volume between groups, a greater plasmacrit in anaemics would suggest a greater total plasma volume. Since lactate is carried in the plasma, anaemics would appear to have a greater dilution of lactate and thus a greater lactate production when compared to non-anaemics. Also, as Figure 9B shows, at any level of relative work red cell 2,3 diphosphoglycerate (2,3 DPG) concentration is lower in anaemics. This pattern is contradictory to studies of anaemia at low altitude where increased 2,3 DPG have been associated with anaemia and a right-shifted O_2-haemoglobin dissociation curve (Finch & Lenfant, 1972). Studies of normal subjects at high

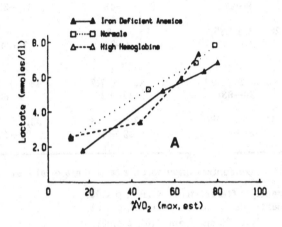

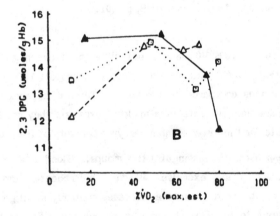

Figure 9:

Metabolic responses during submaximal exercise relative to percentage of estimated maximum oxygen consumption (% VO_2 max est.) in anaemic, normal and high haemoglobin male subjects at high altitude (3600m).

A: Blood lactate:
B: 2,3 diphosphorylate (2,3 DPG)

a = significantly different from normal, p<.05, ANOVA;
b = significantly different from high haemoglobin, p<.05, ANOVA.

altitude also report increased 2,3 DPG compared to low altitude (Eaton et al,1969; Arnaud et al, 1976; Clench et al, 1982) and indicate a right-shifted dissociation curve.   It appears that extreme tissue hypoxia associated with anaemia at moderately high altitudes results in a left-shift in the O_2-haemo-globin dissociation curve.   Considering the alterations in ventilation, reduced D_Lco and apparent lack of arterial desaturations among the iron-deficient anaemics during exercise, it might be suggested that the left-shifted curve is favouring oxygen loading at the lungs and reducing the potential for tissue unloading under these extreme hypoxic conditions.   In a theoretical paper, Benkowitz et al (1982) suggest that adaptation to hypoxia resulting from limitations in pulmonary diffusion would be enhanced by a left-shift in the dissociation curve.   Also, anaemia at moderately high altitude is not unlike the conditions of hypobaric hypoxia experienced by non-anaemics above 6000 meters where a left-shifted dissociation curve has been shown to be adaptive in rats (Eaton et al, 1974).   Moreover, humans with high O_2-haemoglobin affinity have been shown to be well-adapted at high altitudes (3100 m) (Hebbel et al, 1978).

CONCLUSIONS

In this research we have been able to demonstrate that iron-deficiency exists in the high altitude population of La Paz, Bolivia, and that anaemia is associated almost exclusively with iron deficiency in this population. However, given the amount of iron deficiency observed we are surprised at the relatively low prevalence of anaemia.

Using mixed distribution analysis, it was possible to model the haemo-globin distributions for the various subpopulations of anaemic, normal and polycythaemic subjects and to derive cut-off values for haemoglobin that can identify these groups.   These cut-off values seem to discriminate subgroups reasonably well in terms of their physical work capacities and response to steady-state submaximal exercise.   This is particularly true of anaemics whose work capacities are seventeen to twenty-one percent lower than non-anaemics in spite of ventilatory and metabolic adaptations that raise P_AO_2, P_aO_2, and S_aO_2, but do not bring C_aO_2 and systemic oxygen transport up to non-anaemic levels. Therefore, iron deficiency adds a significant physiologic burden to the individual at high altitudes: a burden not unlike exposure to even higher altitudes than the ones where humans currently reside.   This exaggerated hypoxia places real limits on productivity and other aspects of physical performance.

REFERENCES

Arnaud, J., Vergnes, H. & Gutierrez, N. (1976).   Funcion respiratoria y metabolismo del eritrocito en altitudes extremes.   Hematologia,   **22–23**, 73–87.

Baker, P.T. (1969).   Human adaptation to high altitude.   Science   **163**, 1149–1156.

Balke, B. & Ware, R.W. (1959).   An experimental study of 'physical fitness' of Ai Force Personnel.   U.S. Armed Forces Medical Journal,   X (6), 625–688.

Basta, S.S., Soekirman, K.D. & Scrimshaw, N. (1979).   Iron deficiency anemia and the productivity of adult males in Indonesia.   American Journal of Cl:-ical Nutrition.   **32**, 916–925.

Beall, C.M., Brittenham, G.M. & Kingman, P.S. (1983).   Reappraisal of Andean high altitude erythrocytosis from a Himalayan perspective.   Seminar in Respiratory Medicine,   **5**, 195–201.

Beall, C.M. & Reichsman, A.B. (1984).   Hemoglobin levels in a Himalayan high altitude population.   American Journal of Physical Anthropology,   **63**, 301–306.

Benkowitz, H.Z., Wagner, D.D. & West, J. (1982).   Effects of change in P_{50} on exercise tolerance at high altitude: a theoretical study.   Journal of Applied Physiology.

Buskirk, E.R. (1978).   Work capacity of high altitude natives,   In: P.T. Baker (ed.),   The Biology ^f High Altitude Peoples, pp. 173–187.   Cambridge: Cambridge University Press.

Cook, J.D., Alvarado, J. & Gutnisky, A. (1971).   Nutritional deficiency and anemia in Latin America: a collaborative study.   Blood,   **38**, 591–603.

Cook, J.D. & Finch, C.A. (1979).   Assessing iron status of a population.   American Journal of Clinical Nutrition,   **32**, 2115–2119.

Cook, J.D., Finch, C.A. & Smith, N.J. (1976).   Evaluation of the iron status of a population.   Blood,   **48**, 449–455.

Dallman, P.R. (1982).   Iron deficiency: distinguishing the effects of anemia from muscle iron deficiency on work performance,   In: P. Saltman & J. Hegenauer (eds.),   The Biochemistry and Physiology of Iron, pp. 509–523.   Elsevier, North Holland.

Dallman, P.R., Siimes, M.A. & Stekel, A. (1980).   Iron-deficiency in infancy and childhood.   America⁻ Journal of Clinical Nutrition,   **33**, 118–860.

Davies, C.T.M., Chukweumeka, A.C. & Van Haaren, J.P.M. (1973).   Iron-deficiency anemia: its effect on maximum aerobic power and responses to exercise in African males aged 17–40 years.   Cl:nical Science,   **44**, 555–562.

Dinkara, P., Blumenthal, W.S., J⁻hnston, R.F., Kaufman, L.A. & Solnick, P.N. (1970).   The effect of anemia on pulmonary diffusing capacity with derivation of a correction equation.   American Review of Respiratory Disease,   **102**, 965–969.

Eaton, J.W., Brewer, G.J. & Grover, R.F. (1969).   The role of red cell 2,3-diphosphoglycerate in the adaptation of humans to high altitude.   Journ⁻l of Laboratory Cl:⁻ical Medicine,   **78**, 603–609.

Eaton, J., Skelton, T.D. & Berger, E. (1974).   Survival at extreme altitude: protective effect of increasing hemoglobin oxygen affinity.   Science, **183**, 43-744.

Finch, C.A., Golinick, P.D. & Mackler, B. (1979).   Lactic acidosis as a result of iron-deficiency.   Journal of Clinical Investigation, **64**, 129-137.

Finch, C.A. & Lenfant, C. (1972).   Oxygen transport in man.   New England Journal of Medicine, **286**, 407-415.

Finch, C., Miller, L., Inamdar, A., Person, R., Seilar, K. & Mackler, B. (1976).   Iron deficiency in the rat.   Journal of Clinical Investigation, **58**, 447-453.

Frisancho, A.R. (1976).   Growth and morphology at high altitude,   In: P.T. Baker & M.A. Little (eds.),   Man in the Andes: a multidisciplinary study of high altitude Quechua.   Stroudsberg, Philadelphia: Dowden, Hutchinson & Ross.

Frisancho, A.R., Martinez, C., Velasuez, T., Sanchez, J. & Montoye, H. (1973).   Influence of developmental adaptation on aerobic capacity at high altitude.   Journal of Applied Physiology, **34**, 176-180.

Galen, R.S. & Gambino, S.R. (1975).   Beyond Normality: the predictive value and efficiency of medical diagnosis.   New York: Wiley.

Gardner, G.W., Edgerton, V.R., Senewiratne, V., Barnard, R.J. & Ohira, Y. (1977).   Physical work capacity and metabolic stress in subjects with iron-deficiency anemia.   American Journal of Clinical Nutrition,   **30**, 910-918.

Garby, L., Irnell, L. & Werner, I. (1969).   Iron deficiency in women of fertile age in a Swedish community.   II.   Efficiency of several laboratory tests to predict response to iron supplementation.   III. Estimation of prevalence based on response to iron supplementation.   Acta Medica Scandinavica, **135**,   107-117.

Garruto, R. (1976).   Hematology,   In: P.T. Baker & M.A. Little (eds.),   Man in the Andes, pp. 261-282.   Stroudsburg, Philadelphia: Dowden, Hutchinson & Ross.

Garruto, R.M. & Dutt, J.S. (1983).   Lack of prominent compensatory polycythemia in traditional native Andeans living at 4200 meters.   American Journal of physical Anthropology, **61**, 355-366.

Greksa, L.P., Haas, J.D., Leatherman, T.L. & Thomas, R.B. (1984).   Submaximal and maximal work capacity of high altitude Aymara males.   Annals of Human Biology, **11**, 227-234.

Habicht, J.-P., Meyers, L.D. & Brownie, C. (1982).   Indicators for identifying and counting the improperly nourished.   American Journal of Clinical Nutrition, **35**, 1241-1254.

Habicht, J.-P. (1980).   Some characteristics of indicators of nutritional status for use in screening and surveillance.   American Journal of Clinical Nutrition, **33**, 531-535.

Hebbel, R.P., Eaton, J.W., Kronenberg, R.S., Zanjani, E.D., Moore, L.G. & Berger, E.M. (1978).   Human Llamas: adaptation to altitude in subjects with high hemoglobin oxygen affinity.   Journal of Clinical Investigation, **62**, 593-600.

Hurtado, A. & Guzman-Barron, A. (1934).   Estudios antropometricos en 100 indios peruanos de los departementos centrales del Peru.   Revista Sanidad Militar (Lima), **7**, 113-138.

Hurtado, A., Merino, C. & Delgado, E. (1945). Influence of anoxemia n the hemopoietic activity. Archives of Internal Medicine, **75**, 284-323.

ICNND (1964). Bolivia: Nutrition Survey. Interdepartmental Committee on Nutrition for National Defense. Washington, D.C.: U.S. Department of Defence.

Kollias, J., Buskirk, E.R., Akers, R.F., Prokop, E.K., Baker, P.T. & Picon-Reategui, E. (1968). Work capacity of long time residents and newcomer to altitude. Journal of Applied Physiology, **24**, 792-799.

Leatherman, T.L., Greksa, L.P., Haas, J.D. & Thomas, R.B. (1984). Anthropometric survey of high altitude Bolivian porters. Annals of Human Biology, **11**, 253-256.

Mazess, R.B. (1969). Exercise performance at high altitude in Peru. Federation Proceedings, **28**, 1301-1306.

McArdle, W., Katch, F. & Katch, V. (1981). Exercise Physiology: energy, nutrition and human performance. Philadelphia, Pennsylvania: Lea & Febiger.

Meyers, L.D., Habicht, J.-P., Johnson, C.L. & Brownie, C. (1983). Prevalences of anemia and iron deficiency anemia in black and white women in the United States estimated by two methods. American Journal of Public Health, **73**, 1042-1049.

Mueller, W.H., Murillo, F., Palamino, H., Badzioch, M., Chakraborty, R., Fuerst, P. & Schull, W.J. (1980). The Aymara of western Bolivia: V. Growth and development in an hypoxic environment. Human Biology, **52**, 529-546.

Quinn, V.J. (1982). The relationship between hemoglobin and iron status in primary school children living at high altitude in La Paz, Bolivia. Unpublished MS thesis, Cornell University, Ithaca, N.Y.

Quevado, A.S. (1961). Anthropologia del indigena cuzqueno. Reuista del Universidad (Cuzco), **120**, 159-270.

Roach, R.C. (1987). Physiological responses to steady state work in man with variable hemoglobin and iron status at high altitude. Unpublished MS thesi , Cornell University, Ithaca, New York.

Sacchetti, A. (1964). Capacidad respiratoria y aclimatacion en las razas andinas. Ensayo de anthropologia fisco-amxologica. Journada de Sociedad Americanistas, **53**, 9-83.

Sauberlich, H.E., Dowdy, R.P. & Skala, J.H. (1974). Laboratory Tests for the Assessment of Nutritional Status. Cleveland, Ohio: CRC Press.

Spurr, G.B. (1983). Nutritional status and physical work capacity. Yearbook of Physical Anthropology, **26**, 1-35.

Tufts, D.A., Haas, J.D., Beard, J.L. & Spielvogel, H. (1985). Distribution of hemoglobin and functional consequences of anemia in adult males at high altitude. American Journal of Clinical Nutrition, **42**, 1-11.

U.S. AID (1979). Proyecto de Anemia. U.S. Agency for International Development, La Paz, Bolivia. Unpublished report.

World Health Organization (1968). Nutritional Anemias. World Health Organization Technical Report Series, No. 405. Geneva: WHO.

MARGINAL MALNUTRITION IN CHILDHOOD: IMPLICATIONS FOR ADULT WORK CAPACITY AND PRODUCTIVITY

G. B. SPURR

Department of Physiology, Medical College of Wisconsin, Milwaukee, Wisconsin, U.S.A.

INTRODUCTION

In those areas of the world which lie between the Tropics of Cancer and Capricorn, the great majority of the countries can be categorised as developing nations, where mechanisation is at a minimum and economic productivity relies heavily on human labour (Smil, 1979). In a recent listing of some 39 countries having significant deficits in the daily caloric intake of their populations, 30 countries were located in the tropics (Berg, 1981). Consequently, any discussion of physical work capacity of individuals living in the tropics must take into account the effects of undernutrition and poor nutritional status.

From data published by the United Nations (1980), it is possible to estimate the percentage of the economically active populations engaged in moderate to heavy physical work in some tropical countries (Table 1). Estimates of the number of people who are chronically undernourished vary from one in four to one in eight of the world's population (Soedjatmoko, 1981). The two-fold difference in these values reflects the difficulty of estimating the prevalence of undernutrition, but even the lowest estimate indicates the magnitude of the problem. If a direct relationship exists between nutritional status and physical work capacity and, in turn, between the latter and productivity in moderate to heavy work, the data in Table 1 would indicate a retarding effect of undernutrition on the economic development of affected populations (Berg, 1973).

In the case of children, while the prevalence of severe malnutrition has been estimated at 1 to 3 percent of the population, Bengoa and Denoso (1974) concluded that at least 10 times that figure have less severe forms. The proportion of children who pass through longer or shorter periods of undernutrition severe enough to produce a slowing of growth is no doubt much higher in the

Table 1:   Percentages of economically active
populations engaged in moderate to heavy
physical work (agriculture, hunting, fishing,
forestry, mining and construction) of some
tropical countries (U.N. Demographic Year
Book, 1980)

Country	Male	Female
Honduras	75.1	7.5
Ecuador	58.7	13.0
Guatemala	70.5	7.1
Brazil	57.9	20.7
Costa Rica	52.7	4.3
Venezuela	33.9	3.6
Sri Lanka	50.8	62.1
Phillipines	72.0	35.8
Cameroon	67.6	87.4

developing countries of the world.   Reddy (1981) states that on the basis of
growth failure, it is estimated that 80 to 90% of the pre-school children in poor
segments of the population are malnourished.

The present report is concerned with the effects on physical work
capacity of moderate malnutrition during growth and their implications for work
productivity as adults.   It will review briefly the effects of chronic, naturally
occurring malnutrition on the work capacity of adults and the relationship
between work capacity and productivity in physically demanding work tasks.
The effects of marginal malnutrition on growth and work capacity of boys 6-16
years of age will be limited to three categories;   anthropometric and
developmental data of normal and marginally undernourished boys, the growth of
physical work capacity, and the mechanical efficiency of submaximal work in
these children.

THE STUDY SITE

The studies from our laboratory were carried out in Cali, Colombia, and
its rural environs.   Colombia is a country considered to be at a middle level of
development.   Cali is an industrial city of 1.7 million inhabitants, the third
largest in Colombia, located 3° 22' north of the equator at an altitude of 976
meters.   In common with other Latin American cities, during the past 20-25
years it has undergone rapid growth due, in part, to an influx of population from
rural areas.   It enjoys a year-round average temperature of some 24°C (high
29°C, low 18°C) which varies little throughout the year so that wide seasonal

differences in ambient temperature were not a factor in the studies to be described. There are two "rainy" seasons (March to June and October to December) during which average monthly rainfall may reach a maximum of 18 cm, while maximum rainfall during the "dry" seasons is 6-7 cm/month (CVC, 1984).

STUDIES IN ADULTS

The results of studies of up to 10 days of acute starvation (Consolazio et al, 1967), or up to 24 weeks of semi-starvation (Keys et al, 1950), under laboratory conditions have been reviewed recently (Spurr, 1983) and will not be repeated here. Instead, attention will be directed to the few studies available in the literature on the relationships among nutritional status, physical work capacity (PWC) as measured by the maximal oxygen consumption ($\dot{V}O_2$ max) and productivity in some work situations, with comments on endurance as a component of productivity in chronically undernourished subjects. Since the malnourished individual is usually not working (a reason for his malnourished state), particularly in moderate or heavy work tasks, it has not been possible to relate malnourished states directly to productivity. Rather, the attempt has been made to relate both nutritional status and productivity (measured in nutritionally normal, employed subjects to a common measurement ($\dot{V}O_2$ max) and from these relationships to infer the association between nutritional status and productivity in moderate to heavy work. Much of what follows is a synthesis of research from our laboratory, although many other investigators have made important contributions to current knowledge. Probably because of the nature of the studies, most reports in the literature are the result of measurements in male subjects, and unless indicated otherwise, this assumption is to be made.

Malnutrition and physical work capacity ($\dot{V}O_2$ max)

Viteri (1971) compared the PWC of several groups of young Guatemalan adults, one of which, their subjects from San Antonio La Paz (S.A.P.), can probably be considered at least marginally malnourished, on the basis of their adiposity, lean body mass (LBM), and muscle cell mass (MCM, calculated from daily creatinine excretion). These are presumably the same (unsupplemented) subjects discussed by them in their comparisons of supplemented and unsupplemented agricultural workers (Viteri & Toruń, 1975). The S.A.P. group, another group of recent inductees into the army who were from a similar rural socioeconomic background, and ten of the supplemented agricultural workers discussed above, all had significantly lower $\dot{V}O_2$ max and maximal aerobic power (expressed as per kg of body weight or LBM) than army cadets from middle or

upper socioeconomic levels who had never been exposed to nutritional deprivation. When compared on the basis of "cell residue" (body weight less fat, water
and bone mineral; Allen et al, 1959), all differences in maximal aerobic power
between groups disappeared. Viteri (1971) made the important observation that
the differences in maximal aerobic power were due to differences in body
composition and not to differences in cell function.

   We have studied three groups of chronically malnourished adult males who
were selected for their existing degree of undernutrition (Barac-Nieto et al,
1978a). The most severely malnourished of these subjects were also studied
during a 45-day basal period in the hopsital and during 79 days of a dietary
repletion regime (Barac-Nieto et al, 1979,1980). Subjects were classified into
those with mild (M), intermediate (I) and severe (S) malnutrition based on their
weight/height (W/H) ratios, serum albumin (AL) concentrations and daily
creatinine excretions per meter of height (Cr/H) as detailed in Table 2. Each
group was significantly different (p < 0.001) from the other two in regard to each
variable used in the classification. Detailed body composition and biochemical
measurements of the three groups were made shortly after admission to the
hospital metabolic ward (Barac-Nieto et al, 1978b) and during the dietary
repletion regime of Group S (Barac-Nieto et al, 1979). Upon entry into the
hospital, the subjects were placed on an energy intake (2240 Kcal/day) adequate
for the sedentary conditions of the metabolic ward but were maintained on the
same protein intake (27g/day) as they were ingesting prior to entry. Studies of
work capacity and endurance in the severely malnourished men were made at the
begining and end of the 45-day basal period on this diet. The protein intake was
then increased to 100 g/day for the 79-day repletion regime; the increased
caloric intake from proteins was balanced by reducing carbohydrate intake to
maintain the diets isocaloric. Measurements of $\dot{V}O_2$ max and endurance were
repeated at 90 and 124 hospital days. The results for the three groups and the
changes in the severely malnourished men during dietary repletion are presented
in Figures 1 and 2 and compared with data on 107 nutritionally normal control
subjects who were sugar cane cutters (Spurr et al, 1975), loaders (Spurr et al,
1977a) or general agricultural workers (Maksud et al, 1976). There were
progressive differences in body weight, weight/height ratio, serum albumin and
total proteins in the control (C), M, I and S groups (Fig. 1). Groups C and M
were not significantly different in regard to hematocrit and blood hemoglobin
but I and S were significantly and progressively depressed in these measurements. There was a slight gain in body weight of Group S during the basal period
but otherwise, the variables did not change. Weight, W/H ratio, and the serum

Table 2: Selection criteria, means and standard deviations of mild (M), intermediate (I) and severely (S) malnourished adult males

Subject Groups	Weight/Height kg/m	Serum Albumin g/dl	Daily Creatinine/Height mg/day/m
M (n=11)	>32 33.3 ± 2.1	>3.5 3.8 ± 0.5	>600 660 ± 67
I (n=18)	29 – 32 30.8 ± 2.0	2.5 – 3.5 3.0 ± 0.7	450 – 600 559 ± 75
S (n=18)	<29 27.4 ± 2.1	<2.5 2.1 ± 0.5	<450 391 ± 76

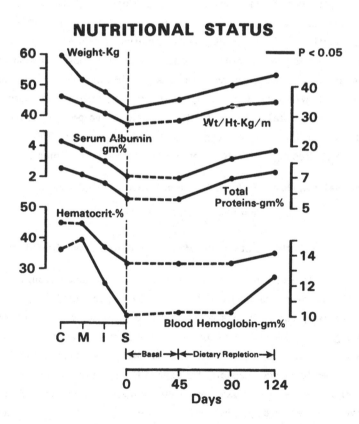

Figure 1: Mean weight, weight/height and blood variables in nuritionally normal (C) men and those with mild (M), intermediate (I), and severe (S) malnutrition. The severely undernourished were studied during a basal period on adequate calories and low protein followed by a dietary repletion period on an isocaloric but high protein diet. Solid lines connect points which are significantly different from each other. (From Spurr, 1983.)

proteins showed progressive improvement during the repletion regime but the hematological values did not show improvement until the final round of measurements (Fig. 1).

Figure 2 presents the results for maximal heart rate (f_H max), maximal aerobic power ($\dot{V}O_2$ max/kg body weight) and $\dot{V}O_2$ max (1/min) for the control and malnourished subjects.   Average max f_H values were not different in the various groups nor did they change during dietary repletion.   However, $\dot{V}O_2$ max and maximal aerobic power were progressively less in C, M, I and S subjects, did not change in the latter during the basal period, and then progressively improved during dietary repletion although they did not return to the level of Group M during the period of study.   Figure 2 also expresses a theoretical submaximal work load of 0.75 1/min $\dot{V}O_2$ in terms of % $\dot{V}O_2$ max for each of the groups.   It is clear that $\dot{V}O_2$ max and maximal aerobic power are markedly depressed in chronic malnutrition and that the degree of reduction is related to the severity of depression in nutritional status.   Using the three groups of malnourished subjects, a stepwise multiple regression analysis (Barac-Nieto et al, 1978a) revealed that the W/H ratio (kg/m), log of the sum of triceps and subscapular skinfolds in mm (Sk), total body Hb (TotHb) obtained as the product of blood Hb and blood volume (g/kg body weight), and daily creatinine (Cr) excretion (g/day/kg) contributed significantly to the variation in $\dot{V}O_2$ max (1/min);

$$\dot{V}O_2 \text{ max} = 0.095 \text{ W/H} - 0.152 \text{ Sk} + 0.087 \text{ TotHb} + 0.031 \text{ Cr} - 2.550 \qquad (1)$$

$$(r = 0.931; \text{ S.E.E.} = 0.21)$$

All of the variables in the equation are, of course, related to nutritional status.

Figure 3 expresses the data for the three malnourished groups and for Group S during recovery, in terms of various body compartments.   It was not possible to do body composition studies on the control subjects.   The salient feature of Figure 3 is that over 80% of the difference in $\dot{V}O_2$ max between M and S subjects is accounted for by difference in MCM.   The remaining difference might be ascribed to reduced capacity for oxygen transport either because of low blood Hb (Fig. 1) or reduced maximum cardiac output.   A number of investigators have reported changes in cardiac muscle in nutritional deficiencies (Gillanders, 1951; Correa et al, 1963; Araujo et al, 1970) which could result in decreases in the force of contraction and perhaps therefore in maximal stroke volume.   However, there do not seem to be any reports of studies on maximum cardiac output in malnourished subjects.   Another possibility is that the skeletal muscle cells have reduced maximal aerobic power because of reduced oxidative

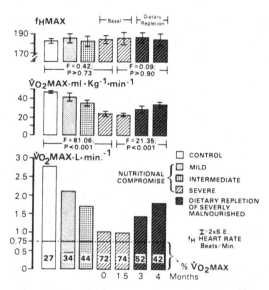

Figure 2:   Maximum heart rates (F_H), aerobic power and $\dot{V}O_2$ max in control, undernourished and severely malnourished subjects and during dietary repletion of the latter.   Lower panel shows a fixed sub-maximal work load ($0.75\ l\cdot min^{-1}$) in terms of % $\dot{V}O_2$ max.   (From Spurr, 1983.)

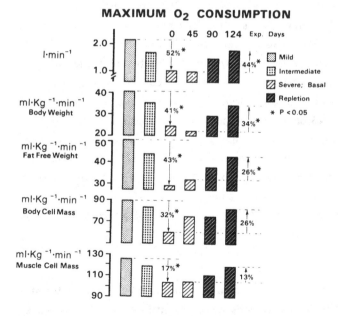

Figure 3: $\dot{V}O_2$ max expressed in terms of various body compartments for the undernourished subjects and during dietary repletion of the most severely malnourished.   (From Spurr, 1983.)

enzyme content.   Tasker and Tulpule (1964) found a marked decrease in the activities of oxidative enzymes in skeletal muscle of protein deficient rats, and Raju (1974) reported that after recovery from 13 weeks of reduced protein intake, rat skeletal muscle had an increase in glycolytic and a decrease in oxydative enzymes and activity.   However, there are no studies which have measured similar biochemical changes in humans, although Lopes et al (1982) have recently shown that malnourished patients exhibited marked impairment in muscle function.   There were both an increased muscle fatigability in static muscular contraction and a changed pattern of muscle contraction and relaxation which were reversed in patients undergoing nutritional supplementation.   Their data indicate the possibility of a decreased content of ATP and phosphocreatine in the skeletal muscle tissue of malnourished subjects.   The data of Heymsfield et al (1982) indicate changes in the biochemical composition of skeletal muscle in both acute and chronic semistarvation, particularly in glycogen and total energy contents.   In any event, it should be emphasised that the $\dot{V}O_2$ max not accounted for by differences in MCM is small (Fig. 3).   After two and a half months of recovery the $\dot{V}O_2$ max increased significantly in l/min and when expressed in terms of body weight and LBM, but, although mean values were elevated in terms of body cell mass (BCM) and MCM, the increases were not statistically significant.   However, at the termination of the experiment, physical work capacity (PWC) had not returned to values comparable to those seen in mild malnutrition (Figs. 2 and 3) which indicates that the recovery process is a long one, particularly when carried out under the sedentary conditions of the hospital metabolic ward.   It is interesting to note that the $\dot{V}O_2$ max was increased 45 days after beginning the repletion diet (90 hospital days) when blood Hb concentration had not yet increased (Fig. 1) but MCM was significantly increased over basal values (Barac-Nieto et al, 1979).   This also points to a primary dependence of $\dot{V}O_2$ max on MCM (Fig. 3).   Furthermore, it appears that supplying adequate calories alone was not sufficient to bring about an increase in $\dot{V}O_2$ max or MCM and that only after increasing the protein intake to 100 g/day was there improvement in these two variables.

Endurance

An endurance test is carried out on a treadmill or bicycle ergometer at a work load ($\dot{V}O_2$) of 70-80% of the subject's maximum until exhaustion supervenes, usually with the f_H within about 5 beats of $f_{H\ max}$. Because of the difficulty in performing this test, only a few laboratories have attempted measurement of endurance times in normal individuals and, to our knowledge, none except our own in malnourished subjects.

The relationship between relative work load (% $\dot{V}O_2$ max) and endurance time is a negative exponential one, having the form

$$\% \ \dot{V}O_2 \ max \ = \ Ae^{-Kt} \qquad\qquad (2)$$

where, in the linearised form, A = the intercept, K the slope, and t the endurance time.   From a number of sources, it is known that the maximum relative work load that can be sustained for an 8-hour work day usually does not exceed about 40% $\dot{V}O_2$ max.   Thus Michael et al (1961) found in laboratory treadmill work, that 8 hours could be tolerated without undue fatigue when the relative load did not exceed 35% $\dot{V}O_2$ max.   In the building industry, Åstrand (1967) reported that about 40% $\dot{V}O_2$ max was the upper limit that could be tolerated for an 8-hour work day and we have estimated that sugar cane cutters worked at about 35% of the $\dot{V}O_2$ max during an 8-hour day (Spurr et al, 1975).   These studies were performed in physically fit subjects.   Sedentary individuals can be expected to have lower upper limits for 8 hours of work (Åstrand & Rodahl, 1970, p. 292).

We have measured endurance times in the groups of malnourished subjects described in one section on malnutrition and physical work capacity by two endurance tests carried out at 90-95% and 80-85% $\dot{V}O_2$ max (Barac-Nieto et al, 1978a, 1980).   Using the linearised form of equation 2 and substituting the actual values for % $\dot{V}O_2$ max and endurance times in the two tests, it was possible to calculate the endurance time at 80% $\dot{V}O_2$ max (T_{80}).   We were unable to demonstrate any significant differences between the three groups (M, I and S) of malnourished men; T_{80} averaged 97 ± 12 min (mean ± S.E.) in all subjects.   However, it might be assumed that the $\dot{V}O_2$ max of Group S subjects would be about 2.4 l/min had they not been malnourished, and that about 35% (0.84 l/min) could be sustained for an 8-hour work day.   The value of 0.84 l/min is 80% of the $\dot{V}O_2$ max (1.05 l/min) for these subjects who had maximum endurance times at this relative work load of a little over 1.5 hours, a loss of about 6.5 hours of daily working time or about an 80% reduction in productive potential (Barac-Nieto et al, 1978a).   Using a similar method of estimation, Barac-Nieto (1984) has calculated a 16% reduction in work output of the M subjects, a 35% decrease in I and a 78% reduction in S men.

In the case of Group S during dietary repletion, an interesting change in T_{80} was observed.   Endurance times were significantly reduced from 113 min at the first measurement of the basal period to 42 min at the final determination at the end of the dietary repletion (Barac-Nieto et al, 1980).   The explanation for this surprising reduction is still not clear.   Hanson-Smith et al (1977) reported decreased work endurance times in rats on high protein diets compared to animals ingesting an isocaloric carbohydrate diet and Bergstrom et al (1967) and Collnick et al (1972) have shown that diets in which the energy value of carbohydrate has been replaced with fat and/or protein lead to reduced stores of muscle glycogen. Furthermore, Bergstrom et al (1967) demonstrated that the maximum endurance time in humans is directly related to the initial glycogen content of skeletal muscle. During the dietary repletion period of the Group S subjects, carbohydrate intake was reduced from 64% to 50% of calories. In a normal individual this amount of carbohydrate should be sufficient to maintain muscle glycogen stores, but definitive studies seem not to have been done (Durnin, 1982). The rebuilt muscle tissue of Group S subjects may not store glycogen normally and, together with the lack of regular exercise in the protracted sedentary existence in the metabolic ward, may lead to reduced muscle glycogen and shorter endurance times. Heymsfield et al (1982) found reduced muscle glycogen in subjects who had undergone acute or chronic semistarvation prior to death. The subject of muscle nutritive supply and metabolism and the endocrine responses which regulate them during both short term and prolonged exercise has not been investigated in malnourished individuals. Even though there is little reason at the moment to suspect abnormal muscle function in acute exercise testing to maximum levels, the responses to prolonged exercise may be worth investigating.

Productivity and physical work capacity

Having established a direct relationship between nutritional status and physical work capacity in undernourished men, attention can now be directed towards the association between $\dot{V}O_2$ max and productivity. The amount of work done in terms of output of a product is very difficult to measure, particularly in the lighter work tasks where the intellectual components may have as much or more to do with "productivity" as the physical use of one's body. In moderate and heavy work it has sometimes been possible to estimate productivity by measuring the quantity of product or income where piece-work is the basis for payment of the worker. Sugar cane cutting and loading are heavy work tasks where the weight of cane cut or loaded is measured carefully since workers are usually paid by the tonnage. Because the pay scale in many sugar harvesting operations is very low, one might expect that the motivation factor would be fairly similar in different groups of workers and that they would work close to the limit of their physical capacities. Also, logging is heavy physical work (Durnin & Passmore, 1967) and has been used to relate productivity to worker characteristics. The time to accomplish standard work tasks is another method which has been utilised to estimate productivity.

Hansson (1965) measured submaximal work and estimated $\dot{V}O_2$ max in a group of "top" producing lumberjacks and a group of average producers. There were no differences between the two groups in measurements of height, weight, circumferences of lower and upper arms and chest, and muscle strength in the hands and arms or when vertically lifting with two hands. However, the top producers had a higher estimated $\dot{V}O_2$ max than the average group. Davies (1973b) studied sugar cane cutters in East Africa, dividing them into high, medium and low producers based on the daily tonnage cut. As in the case of the loggers of Hansson (1965), he found no difference in the three groups in height, weight, summed skinfolds, LBM, leg volume or the circumferences of biceps and calf, but did encounter a significant correlation between daily productivity and $\dot{V}O_2$ max ($r = 0.46$; $p < 0.001$). Morrison and Blake (1974) observed that six Australian cutters, whose productivity was much higher than their Rhodesian counterparts, also had a higher estimated $\dot{V}O_2$ max, although the difference disappeared when the data were normalised for body weight. Davies et al (1976) also measured productivity in Sudanese cane cutters during a 3 hr period of continuous cutting, and reported a significant correlation between $\dot{V}O_2$ max and rate (kg/min) of cane cutting ($r = 0.26$; $p < 0.01$).

We have studied nutritionally normal sugar cane workers in Colombia, where the tasks of cutting and loading the cane are performed by separate gangs of men. The former is a self-paced and continuous task, while the loading of cane is discontinuous, depending on the availability of wagons. The cutters were divided into good (Group I), average (Group II), and poor (Group III) producers, depending on the daily tonnage cut. The cutters worked at about 35% of their $\dot{V}O_2$ max during the 8 hr day (Spurr et al, 1975), which is close to the maximum that can be sustained for this period of time (Michael et al, 1961; Åstrand, 1967). In relating various anthropometric measurements and age to productivity, there were statistically significant positive correlations of height, weight and LBM with productivity (Spurr et al, 1977b). The correlations with age and body fat were not significant. Figure 4 summarises the relation of $\dot{V}O_2$ max and maximal aerobic power with productivity, both of which were significantly correlated. A step-wise multiple regression analysis revealed that $\dot{V}O_2$ max (l/min), % body fat (F) and height (H; cm) contributed significantly to the variation in productivity (tons/day) such that

$$\text{Productivity} = 0.81 \, \dot{V}O_2 \text{ max} - 0.14F + 0.03H - 1.962 \qquad (3)$$

$$(r = 0.685; \ p < 0.001)$$

The $\dot{V}O_2$ max and body fat are influenced by present nutritional status (Barac-Nieto et al, 1978; Viteri, 1971) and adult height by past nutritional status during the period of growth (Thomson, 1968).   Equation 3 states simply that those who are presently malnourished (low $\dot{V}O_2$ max) or whose height is stunted because of past undernutrition, are at a disadvantage in terms of ability to produce in cutting sugar cane.   The negative coefficient for % of body fat indicates that there is some advantage to low body fat content.   The relatively low correlation coefficients between productivity and $\dot{V}O_2$ max obtained in our studies and those of others (Davies, 1973b; Davies et al, 1976) preclude the use of regression equations in the prediction of productivity and bring into question the homogeneity of motivation alluded to above.   The results shown in Figure 4 indicate that the more physically fit subjects were better producers.   This was also seen in the measurements made on these subjects in the field under their usual working conditions where the f_H during cutting was lowest in Group I, intermediate in Group II and highest in Group III at same $\dot{V}O_2$ (Spurr et al, 1975).

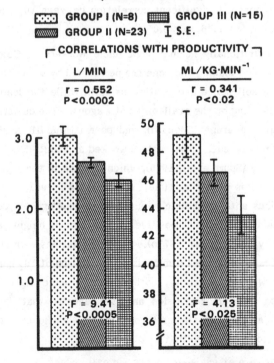

Figure 4: $\dot{V}O_2$ max and maximal aerobic power of good (Group I), average (Group II) and poor (Group III) sugar cane cutters. F ratio values are from a one-way analysis of variance. (From Spurr et al, 1977b.)

Subjects in better physical condition have lower resting and work f_H than non-fit subjects (Åstrand & Rodahl, 1977). There were some men who were able to sustain a % $\dot{V}O_2$ max in excess of 40% for the 8-hour day. Calculations of their efficiency (kg cane cut per 1 of $\dot{V}O_2$) in cutting cane were significantly lower than the others (Spurr et al, 1977c). When these "inefficient" cutters were excluded, there were no significant differences in efficiency in Groups I-III.

Even in the case of the sugar cane loaders, who do not work continuously, productivity was positively correlated with maximal aerobic power and negatively with resting and working f_H, demonstrating again the relationship of productivity to the physical condition of the worker (Spurr et al, 1977a).

In the case of sugar cane cutting, which at an average expenditure of 5 Cal/min per 65 kg of body weight during the 8-hour work day (Spurr et al, 1975) can be classified as moderate industrial work (Durnin & Passmore, 1967), the worker productivity is related to his body size, height, weight and LBM (Spurr et al, 1977b). This has also been demonstrated by Satanarayana et al (1977, 1978) for industrial factory work of presumably less intensity than sugar cane cutting. Their subjects were nutritionally normal workers engaged in the production of detonator fuses which could be measured in terms of the number of fuses produced per day. They found that body weight, height and LBM were significantly correlated with productivity and that after removing the effect of height, weight and LBM, were still significantly correlated with productivity. That is, the total daily work output was significantly higher in those with higher body weight and LBM.

Brooks et al (1979) have shown that the nutritional status of Kenyan road workers was related to productivity. The time to complete three standard work tasks (wheelbarrow work, ditch digging and earth excavation) was related to weight for height as the indicator of nutritional status. The lower the weight for height below the reference population, the longer the time (lower the productivity) for task completion. Also, Wolgemuth et al (1982) have reported that successful dietary supplementation resulted in a statistically significant increase in worker productivity in road building, associated with increases in arm circumference and Hb levels.

STUDIES IN CHILDREN

With the recognition that the reduced work capacity found in malnourished adults was largely the result of reduced muscle mass, the next question was the effect of chronic marginal malnutrition, which is so prevalent in the poorer segments of developing countries, on the growth of work capacity in school-aged children.   There are few studies of exercise and work capacity in malnourished children, and most of these have been carried out using sub-maximal exercise testing.   Areskog et al (1969) determined the physical work capacity at a heart rate of 170 (PWC_{170}) in 10 and 13-year-old Ethiopian boys from public and private schools with the aim of including both poorly nourished and well-nourished subjects.   The older public school boys were shorter, weighed less, and had smaller skinfolds and mid-arm circumferences than their private school counterparts.   The height and weight differences in the younger children were less clear-cut, although mid-arm circumference and skinfolds were smaller in the public school boys, and the average caloric and protein intakes lower in both groups of public school than private school children.   The results of tests of PWC_{170} showed that the performance of the public school (i.e. undernourished) boys was somewhat better than the private school children.   Davies (1973a) predicted $\dot{V}O_2$ max from sub-max bicycle ergometry, demonstrating that malnourished (underweight) children had low values of $\dot{V}O_2$ max, but that maximal aerobic power in terms of body weight, LBM or leg volume were well within the normal range.   Satyanarayana et al (1979) have also reported the results of measurements of PWC_{170} in boys 14-17 years of age categorised according to their nutritional status at age 5 years.   They found that about 64% of the variation in PWC_{170} could be explained by the subjects' body weight at the time of the testing and another 10% by their habitual physical activity levels.   But even severe malnutrition at age 5 had no effect on work performance when they expressed PWC_{170} in terms of body weight.   However, the undernourished subjects had higher values for f_H at the same sub-max work load, i.e. were working at a higher %$\dot{V}O_2$ max than normal children.

In the work to be described from our laboratory, all subjects were boys and had to present their official birth certificate as a first condition of inclusion in the study.   They were grouped into five age groups at two-year intervals from 6 to 16 years of age.   Using the Colombian standards established by Rueda-Williamson et al (1969), children were selected who had weight for age and weight for height between 95 and 110% of predicted, as being nutritionally normal (N) and without a history of undernutrition.   Those with a weight for age <95% but weight for height >95% of the standard were called the low weight for age (W-A) group (stunted) and were considered to have a history of nutritional deprivation but to be nutritionally normal at the time of the study.   A third nutritional group was formed from those children with both weight for age and height >95% of the standard.   These children were considered to be under-nourished at the time of study and are referred to as the low weight for height

(W-H) group (wasted). The reason for choosing 95% as the cut-off point was entirely arbitrary with the expectation that the group averages would be considerably below this point.

The children were also recruited according to socioeconomic status: an upper socioeconomic group (UU) recruited from three of Cali's private schools with high tuition costs, a second lower socioeconomic group from public schools located in several of the poorest city barrios(LU), and a third lower socio-economic group from three rural villages located outside Cali (LR). Conse-quently, each of the two lower socioeconomic groups (LU, LR) had three nutritional groups each (N, W-A, W-H) while the upper socioeconomic urban boys had only N subjects. The numbers of subjects in each of the 35 groups so formed varied between 14 and 60, with most of the groups having approximately 30 subjects each. The total number of boys studied was 1,108. The details of the selection process and the methods employed in the anthropometric and matura-·ion (Spurr et al, 1983b), $\dot{V}O_2$ max (Spurr et al, 1983a), body composition (Barac-Nieto et al, 1984) and work efficiency measurements (Spurr et al, 1984) have been described previously.

Anthropometry and maturation

The average values of weight and height for age, plotted on the U.S. National Center for Health Statistics (NCHS) percentile grids for the 35 groups of boys, are shown in Figure 5.   There were no statistically significant differences in weight for age among the three N groups, which followed the 50th percentile in the younger boys and deviated towards the 25th percentile in the older groups.   The two nutritionally deprived groups were at or below the 5th percentile throughout the age range studied, with the W-H group having significantly lower weights than the W-A boys. The percentages varied between 83 to <70% of the 50th percentile.   There was no significant pattern of differences in weight or height between urban and rural boys.

The heights of the N groups showed significant differences between the UU and the LU and LR groups probably as a result of the higher number of mestizo children in the latter two groups.   When only mestizos of the UU group were included in the analysis, differences disappear, which is in accord with the findings of shorter stature of Mexican-American boys than their Anglo counter-parts (Malina, 1973; Zavaleta & Malina, 1980).   A cardinal feature of the selection process was that the W-H boys were significantly taller than the W-A children. The group percentages of the latter group varied from 92% to <85% of the 50th percentile in the older age groups.

The sum of three skinfold thicknesses, as an indicator of subcutaneous fat, and the mid-arm circumference, are shown in Figure 6. There were no significant differences in the summed skinfold thicknesses between the urban N groups, but W-A and W-H groups were significantly less than N and W-H less than W-A children in both urban and rural groups. Also, the rural children had significantly lower values than their urban counterparts. Another feature of the skinfold data in Figure 6 is that the four nutritionally deprived groups do not show the fall in skinfold thickness associated with puberty in boys (Pařízková, 1977). With the exception of a lack of difference between urban and rural boys, the statistical differences of the mid-arm circumference data follow the same pattern as that described for the summed skinfold data (Figure 6) indicating a deficit of muscle mass in the nutritionally deprived boys (Barac-Nieto et al, 1984).

Tanner (1976) has suggested that since the first thing that happens in the undernourished child is a slowing down of growth, it is relevant to monitor growth velocity. The velocities of growth in weight and height for 53% of the total population of boys is shown in Figure 7 (Spurr et al, 1983b). Since there were no significant socioeconomic differences, the data were combined into the three nutritional groups. The statistical analysis (two-way ANOVA) demonstrated a significant shift of the weight velocity curve to the right in the two nutritionally deprived groups without significant differences between the W-A and W-H groups in either height and weight velocities. Height velocity was also shifted to the right in both deprived groups, but the difference was not statistically significant ($P < 0.11$). However, the age at peak velocity for both weight and height was progressively and significantly retarded in the N, W-A and W-H groups. Furthermore, a frequency analysis of Tanner Scores for sexual maturation (Tanner, 1961) showed that in the 12 to 13.9 and 14 to 16-year-old boys, there was a statistically significant slowing of sexual maturation in the W-A and W-H groups when compared to N boys, without significant difference between deprived groups.

Consequently, the selection process resulted in the separation of boys who were smaller (W-A) and thinner (W-H) than N boys (of whatever socioeconomic background) who were following an essentially normal growth development when compared to either national or international (NCHS; Fig. 5) norms. However, the physiologic data (growth velocity, sexual maturation and lack of fall in skinfold thickness in the older W-A and W-H groups) make it clear that the *reason* for the smallness and thinness is that they are undergoing a process of chronic malnutrition which is no doubt "marginal" in nature but nevertheless real.

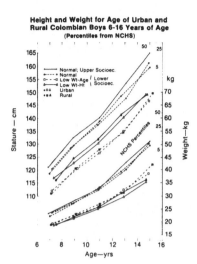

Figure 5: Average values of height and weight plotted as a function of average group ages on U.S. NCHS percentiles. SEM varied between 0.6 and 1.6 cm for height, and 0.2 and 0.9 kg for weight. (From Spurr et al, 1983b.)

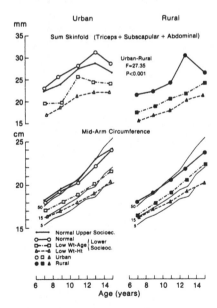

Figure 6: Averaged summed skinfolds and mid-upper arm circumference of urban and rural, upper and lower socioeconomic group boys in three nutritional groups by average group age. The percentile lines for mid-arm circumference are from Frisancho (1974). (From Spurr et al, 1983b.)

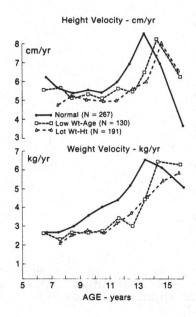

Figure 7:   Average growth velocities of nutritionally normal, low weight for age, and low weight for height boys plotted at the average age midway between two measurements obtained 6-12 months apart. (From Spurr et al, 1983b.)

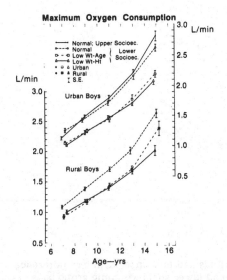

Figure 8:   V̇O$_2$ max of nutritionally normal, low weight for age (Wt-Age) and height (Wt-Ht) urban (right scale) and rural (left scale) boys. Plotted as a function of average group ages. (From Spurr et al, 1983a.)

Furthermore, the fact that there is progressive deviation from predicted values of height and weight for age from younger to older boys (Fig. 5; Spurr et al, 1983b) indicates that the process is cumulative with age.

Growth of work capacity ($\dot{V}O_2$ max)

The growth of $\dot{V}O_2$ max (l/min) in 1,013 of the children described above is shown in Figure 8. There were no significant differences between the N-UU and N-LU boys nor between the urban, W-A, and W-H children, but the $\dot{V}O_2$ max of the nutritionally deprived boys was significantly lower (~85%) than the N children throughout the age range studied. A similar pattern is seen in the LR subjects; W-A and W-H boys have $\dot{V}O_2$ max values which were not different from each other (except in the oldest age group), but which were significantly lower than the N-LR subjects. This last observation can be seen in Figure 9 where maximal aerobic power is presented together with the results of the two-way analysis of variance (ANOVA; age and socioeconomic effects) of the N boys and the three-way ANOVA (age, urban-rural, and nutritional group effects) of the two lower socioeconomic groups. The mean values ranged from 48.7 to 58.3 ml/kg/min in the youngest UU to the oldest LU, W-A groups respectively, with most group averages being in excess of 50 ml/kg/min. There were no differences between the N-UU and N-LU boys, but the N-LR children have significantly lower values than the others, which is the reason for the significant socio-economic effect in the N children. Among the L groups there was a progressive increase with age, lower values in R than U groups, and with W-A and W-H being higher than N groups; all of these relationships were statistically significant. The tendency for higher maximal aerobic power in W-H than W-A boys was statistically significant in the LU boys (F = 6.16, p = 0.014), but not in the LR children (F = 3.47, p = 0.064). These results make it clear that the lower values of $\dot{V}O_2$ max (l/min) for the nutritionally deprived children are due to their lower body weights. This is essentially the same conclusion reached by Davies (1973a) and Satyanarayana et al (1979). Again, as mentioned earlier for adults, there does not appear to be any basic deficit in muscle function in marginally malnourished children, only in the quantity of muscle available for maximal work. We have deliberately avoided analysing these data on the basis of so-called "developmental" age because such an analysis would tend to obscure the differences seen in Figure 8. After all, the responsibilities of adulthood occur with chronological, not developmental age.

The elevated values for maximal aerobic power in the undernourished boys (Fig. 9) are no doubt, at least partially, a reflection of differences in body composition (Fig. 6). To investigate this possibility we have compared the lean

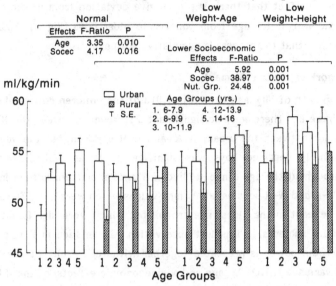

Figure 9:   Maximal aerobic power of nutritionally normal and deprived urban and rural boys with results of 2 and 3-way analysis of variance. (From Spurr et al, 1983a.)

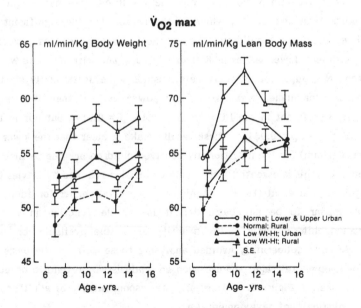

Figure 10:   Aerobic capacities (left panel) and oxidative capacities of the LBM (VO_2 max/LBM)(right panel) at various ages in urban and rural normal and low W-H boys.   (From Barac-Nieto et al, 1984.)

body mass (LBM) estimated from total body water and extracellular fluid volume measurements in sub-groups of our subjects with the LBM obtained by applying the empirical equations of Parizkova (1961) for skinfolds, and have satisfied ourselves that the latter equations developed for European children can be applied to our normal and undernourished Colombian children (Barac-Nieto et al, 1984). To simplify the determination of the contribution of body composition to the results seen in Figure 9, the N-UU and N-LU groups were combined and the analysis limited to the W-H urban and rural deprived groups. The results are presented in Figure 10, where panel A is essentially the same as the results in Figure 9 showing higher values of aerobic capacity in both urban and rural W-H boys than the corresponding N groups. When $\dot{V}O_2$ max is expressed in terms of LBM (Fig. 10B), the differences between rural N and W-H boys largely disappear while the values for urban W-H boys remain significantly higher (except in the youngest age groups) than corresponding N children. The increases in $\dot{V}O_2$ max/LBM in the urban W-H boys could not be accounted for by compositional changes within the LBM: decreases rather than increases in the percentage of cells and muscles within the body seem to occur in the W-H boys (Barac-Nieto et al, 1984). Since aerobic training increases muscle oxidative capacity without increasing muscle mass (Halloszy, 1967), we have interpreted these results to indicate a higher state of physical training of the urban W-H boys, perhaps due to a *relative* increase in their daily physical activity (in relation to their lower total $\dot{V}O_2$ max) in trying to keep up with the activities of their peers, despite their reduced body, lean, cellular and muscle masses (Barac-Nieto et al, 1984). This training effect appeared to be absent in the rural children for unknown reasons perhaps related to weaker peer pressure effects on the W-H boys. The higher $\dot{V}O_2$ max/LBM in urban than rural N boys reinforces the suggestion that urban children have greater access to sports training facilities than rural boys (Shephard et al, 1974; Spurr et al, 1983a).

Implications of $\dot{V}O_2$ max for work and productivity

In order to present more clearly the concepts discussed above, some theorising is necessary to summarise ideas already presented. Figure 11 shows some imaginary data for normally nourished individuals with differing $\dot{V}O_2$ max values for whatever reason: genetic endowment, age, physical condition, etc. Also presented is the theoretical cost of some work task in terms of $\dot{V}O_2$. In the case of subject No. 1, this would represent 50% of his $\dot{V}O_2$ max, 60% of max in subject No. 2, and 75% of max in No. 3. An undernourished subject with severely reduced $\dot{V}O_2$ max (Barac-Nieto et al, 1978) is also known, in whom the work task might be 100% of his $\dot{V}O_2$ max. As already mentioned, up to 35-40%

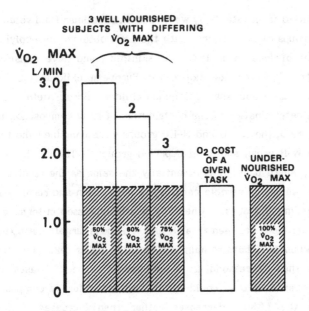

Figure 11:   Theoretical data for three nutritionally normal subjects showing the
load of a given task relative to an undernourished individual.   (From
Spurr, 1983.)

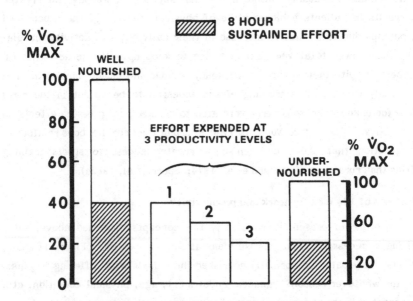

Figure 12:   Theoretical data for a nutritionally normal subject (left scale) and
an undernourished individual (shortened right scale to indicate
reduced total $\dot{V}O_2$ max; see Fig. 2) showing effect on supposed
productivity (no scale) of each subject working at the 40% maximum
which can be sustained for 8 hours. (From Spurr, 1983.)

of the $\dot{V}O_2$ max can be sustained for an 8-hour work day (Michael et al, 1961; Åstrand, 1967; Spurr et al, 1975) in individuals in good physical condition. Figure 12 attempts to show the relationship between the relative load sustained (% $\dot{V}O_2$ max) and productivity in hard physical work. A well-nourished subject willing to work at 40% of his $\dot{V}O_2$ max for 8 hours might have the highest productivity level while the lower productivity levels 2 and 3 would be achieved working at 30 and 20% of $\dot{V}O_2$ max, respectively. An undernourished subject is also shown at the right with a shortened scale of % $\dot{V}O_2$ max to indicate the reduced total $\dot{V}O_2$ max (Fig. 3). Such a person, even utilising the full 40% of his max for the 8 hours of work would be able to produce only at the lowest level. Consequently, the undernourished with reduced working capacity cannot produce as much as the well-nourished. It must be emphasised that these concepts apply only to moderate to heavy physical labour and do not take into account the complexity of the work function and factors such as motivation, the work environment or the advantage that small size might have in some work situations where body translation is involved (Viteri et al, 1981). However, since in the developing world the majority of the adult male work force is engaged in moderate to heavy work (United Nations, 1980; Arteaga, 1976) (Table 1), the implications for economic development are obvious (Berg, 1973). Productivity in the more sedentary occupations has not been studied, if indeed it is measurable.

What do these results portend for the work capacity, and productivity, of these boys when they become adults? Figure 13 presents the $\dot{V}O_2$ max of the four groups of 14-16-year-old urban boys in comparison with the O_2 cost of a work task artificially set at 40% of the N-UU children. In the case of the N-LU children, this amounts to about the same work load (41%), but for the W-A and W-H boys it amounts to 51% and 54% of $\dot{V}O_2$ max, respectively. These latter values are well in excess of the maximum relative loads which can be sustained for an 8-hour work day. Even at 40% of their $\dot{V}O_2$ max, these nutritionally disadvantaged and smaller boys will probably produce, on average, at a lower level than the N boys when they achieve adulthood. The economic implications of allowing a considerable segment of the world's population to grow up as second-rate producers, is evident. Immink et al (1984) have reported results which support the hypothesis that nutritional improvements during childhood are forms of human capital investment which result in higher lifetime earnings in certain occupations and may be associated with upward social and occupational mobility during adulthood. Again, it must be underlined that these considerations apply to a large segment of the population (Table 1) which will become involved in heavy physical work. However, in a larger sense, they also apply to any activity

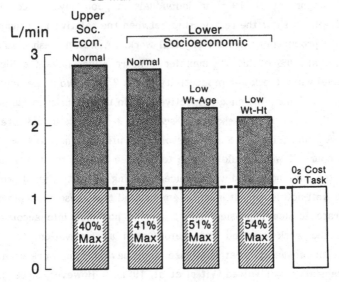

$\dot{V}O_2$ max 14-16 YR OLD BOYS

Figure 13: Average $\dot{V}O_2$ max of the four groups of 14–16-year-old urban boys
showing the relative work load in the lower socioeconomic groups of
a work task selected to be 40% of the $\dot{V}O_2$ max of the upper
socioeconomic group. (From Spurr et al, 1983a.)

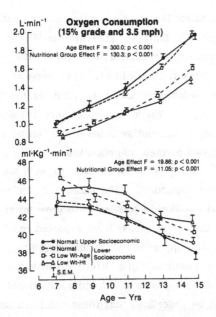

Figure 14: Submaximal $\dot{V}O_2$ at 3.5 mph treadmill speed and 15% grade by
average group ages. (From Spurr et al, 1984.)

which will be more intense relative to the lower max values of the undernourished and therefore more tiring.

EFFICIENCY OF SUB-MAXIMAL WORK IN MALNUTRITION

It has been suggested that adults from developing countries are mechanically more efficient than their taller counterparts from the developed nations (Martorell et al, 1978). The arguments upon which this contention was made have been contested (Spurr, 1983; Spurr et al, 1984). Keys et al (1950) reported measurements of $\dot{V}O_2$ during the last five minutes of a 30-minute walk at 3.5 mph and a 10% grade in 32 adult men before and after a 24-week period of restricted caloric intake during which there was a 24% loss of body weight. With these values and the basal $\dot{V}O_2$ measurements made, it is possible to calculate gross and net efficiency from their data. Gross and net efficiencies before caloric restriction were 17 ± 0.6% and 19.9 ± 0.8% respectively and 17.8 ± 1.2 and 19.9 ± 1.4% after 24 weeks on reduced caloric intake. While the gross efficiency showed a statistically significant increase (paired t = 2.38; p = 0.024) it is difficult to make a case for any physiological significance of such a small change. Net efficiencies were not significantly different.

Edmundson (1979, 1980) has presented data on basal metabolic rate (BMR) and mechanical work efficiency of men in East Java, and together with a review of related literature attempted to make a case for metabolic adaptation to "undernutrition" as judged by the well-known reduction in BMR which occurs in undernourished subjects (Keys et al, 1950) and an increase in mechanical work efficiency in subjects on a chronic low energy intake. The purpose here is to consider only the mechanical work efficiency measurements made on two groups of his subjects who were selected from a group of 54 subjects studied. One group of five subjects was composed of high energy intake individuals (mean = 2754 Kcal/day) who were nearly identical in average height and weight to a group of six low energy intake subjects (mean = 1770 Kcal/day). It would seem,therefore, that the subjects were selected for differences in their existing "metabolic efficiency". The BMR of the high energy intake group was almost twice that of the low energy intake subjects and, perhaps not surprisingly, the gross calculated efficiency at 600 kpm/min on a bicycle ergometer was significantly higher in the latter group than in the former although the difference at 300 kpm/min was not statistically significant. The delta efficiencies calculated from Edmundson's (1979, 1980) data were 16.1 ± 6.48% (range 7.6-23.0%) in the high intake subjects and 23.6 ± 5.26% (range 16.8-

31.0%) in the low intake group and were not significantly different (t = 2.07; p = 0.07).   It may be that a larger number of subjects would show a statistically significant difference in delta efficiency since the range of efficiencies is large. Also, the values are somewhat lower than those reported by others (Gaesser & Brooks, 1975).

We have recently reported on calculations of delta efficiency in a small number of the normal and malnourished males described in the studies on adults above (Spurr & Barac-Nieto, 1984).   There were no statistically significant differences in delta efficiency among the nutritionally normal subjects and the men with mild, intermediate or severe malnutrition.   However, because of the small number of subjects and the retrospective nature of the study, it is difficult to exclude changes in the mechanical efficiency of work as a possible adaptation to chronic malnutrition in adults.   Well-designed prospective studies in adequate numbers of subjects are still needed to answer this question.

We have also reported on the efficiency of submaximal treadmill walking in the normal and marginally malnourished school-aged children described in the section on studies in children above (Spurr et al, 1984).   Because walking speed affects the calculation of work efficiency (Donovan & Brooks, 1977), only those children who walked at 3.5 mph were included in the study.   Since at submaximal work levels there were no detectable urban-rural differences, the lower socioeconomic N, W-A, and W-H groups were combined.   Figure 14 shows the average $\dot{V}O_2$ for the twenty subject groups, in terms of l/min and ml/min/kg body weight, while walking at 15% treadmill grade.   There were no statistically significant differences between N boys nor between the two nutritionally deprived groups in either expression of $\dot{V}O_2$, but N boys had significantly higher values than W-A and W-H children throughout the age range studied.   The reverse situation existed when $\dot{V}O_2$ was expressed in terms of body weight (lower Fig. 14), i.e. decreasing values with age and higher values in the W-A and W-H groups than N groups.

The observation that the O_2 cost (l/min) of treadmill walking or bicycle ergometry increases with age in children and decreases with age when expressed in terms of body weight (ml/kg/min) is an old one (Robinson, 1938; Åstrand, 1951) which has been repeatedly confirmed (Daniels et al, 1978;   Pate, 1981; Montoye, 1982).   The increase in $\dot{V}O_2$ (l/min) with age is an expected result since $\dot{V}O_2$ is related to body mass and older and heavier boys are performing more work at the same treadmill speed and grade.   This effect of body weight would also seem to be a reasonable explanation for the fact that the smaller W-A and W-H children (Fig. 5) follow a similar pattern with age but at lower $\dot{V}O_2$'s than their nutritionally normal counterparts (upper Fig. 14).

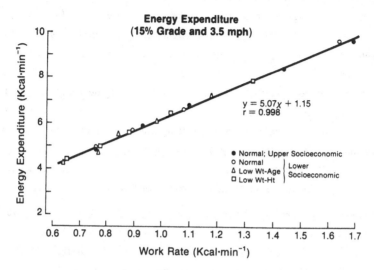

Figure 15: Average energy expenditure of treadmill walking at 15% grade and 3.5 mph. The same symbols are average values of successive age groups; youngest to oldest, left to right. The SEM's of individual points of energy expenditure varied from 0.47 to 1.42. The regression equation is derived from the 20 average group values. (From Spurr et al, 1984.)

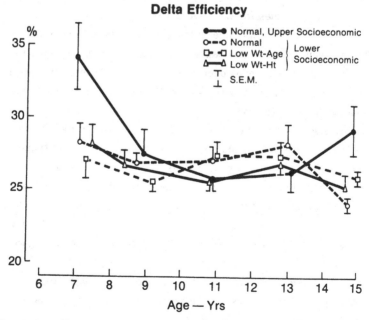

Figure 16: Delta efficiency as a function of average group ages. (From Spurr et al, 1984.)

It has been implied that the fall in submax $\dot{V}O_2$, expressed in terms of body weight, with age represents an improved "efficiency" (Montoye, 1982) perhaps as a result of the fact that younger, shorter individuals need to take more steps at the same speed than older, taller subjects (Åstrand, 1952; Pate, 1981). When Pate (1981) analysed his data on boys and men using stepping frequency as a covariate, the difference in submax $\dot{V}O_2$ between men and boys persisted. However, using only body weight as the covariate, the statistically significant difference disappeared, suggesting that the difference between men and boys may be accounted for by variation in body weight. This would also explain the results reported here (Fig. 5 and lower Fig. 14), i.e. lower O_2 cost per kg of body weight in the heavier (older or nutritionally more adequate) boys.

Indeed, when gross mechanical efficiency was calculate from $\dot{V}O_2$ and work rate in these subjects, the expected results were obtained, viz., increasing values with age and higher values in N boys than the nutritionally deprived children (Spurr et al, 1984). However, Gaesser and Brooks (1975) have pointed out that calculated values of gross efficiency do not agree with steady-state work data, and that estimates of gross efficiency are "artifacts" of the computation. One reason for this is shown in Figure 15, where the average energy expended in treadmill walking at 15% grade and 3.5 mph is plotted against the work accomplished, both expressed as Kcal/min, of all twenty groups of subjects. The values for the W-A and W-H groups fall lower in the line than the two N groups at comparable average ages, but all age, socioeconomic and nutritional groups follow the same linear relationship. The artifact in the calculation of gross efficiency occurs because of the positive intercept value for the linear equation of the relationship in Figure 15. Using this equation, gross efficiency calculated for younger and/or nutritionally deprived (smaller) subjects capable of lower power outputs at the same treadmill grade and walking speed will be lower than that calculated for older and/or more nutritionally adequate children. Gaesser and Brooks (1975) also found in their studies that the delta efficiency ((delta work accomplished/delta energy expended) x 100) did agree with their steady-state work data and concluded that delta efficiency was the most appropriate method of calculating efficiency. The delta efficiencies for the twenty groups of children under discussion are shown in Figure 16. With the exception of the youngest group of UU-N boys (for which we have no explanation) there were no statistically significant age, socioeconomic nor nutritional group differences. The results for nutritionally normal boys have been confirmed by Cooper et al (1984).

Consequently, marginal malnutrition in school-aged Colombian boys does not appear to have any effect on the efficiency of muscular work.   The data from retrospective studies in adult malnutrition of more severe degree is less convincing (Spurr & Barac-Nieto, 1984) and needs further study.   Claims for increased work efficiency in populations of underdeveloped countries (Martorell et al, 1978; Edmundson, 1979, 1980) are unconvincing.

SUMMARY

In considering the physical work capacity of populations living in tropical regions, the nutritional status of the individuals must be taken into account since large numbers of tropical countries can be classified as developing nations in which the incidence of undernutrition may be quite high.   Also, in these countries, relatively high percentages of the economically active populations are engaged in moderate to heavy physical work.   Productivity in some of these occupations is directly related to the physical work capacity of the individual. Furthermore, physical work capacity is dependent on nutritional status such that, relative to the degree of undernutrition, malnourished subjects can have markedly depressed work capacities due largely to decreased muscle mass. By implication, then, the productivity of adult, malnourished persons would also be depressed in those occupations involving moderate to heavy physical work.

During the growth of school-aged children, even marginal malnutrition results in growth retardation, slowing of sexual maturation, delay in the occurrence of the adolescent growth spurt, and reduction in physical work capacity as measured by the $\dot{V}O_2$ max (l/min).   The reduced $\dot{V}O_2$ max appears to be due primarily to the smaller body size (and consequently of the muscle mass) of the affected children.   The results indicate that as adults these children will be able to produce less in those occupations which are physically taxing, and point to the need for investing in adequate nutrition for the economically important segment of the population.

The results to date demonstrate that in the case of marginal malnutrition in young boys, the efficiency of physical work (treadmill walking) is not different from nutritionally normal children.   In the case of more severe malnutrition in adulthood, few data are available, and we do not know if there is a physiological adaptation to chronic malnutrition which might result in an increased efficiency of physical work.   Prospective studies of this question are needed.

ACKNOWLEDGEMENT

The original research reported here from our laboratory has been supported by Contracts Nos. AID/CSD 2943 and AID/TA-C 1424 from the Office of Nutrition of the Agency for International Development, N.I.H. Grant No. HD10814, the Research Service, Wood V.A. Medical Center and the Fundación para la Educación Superior, Cali, Colombia. The work could not have been done without the collaboration of Mario Barac-Nieto, MD, PhD, in all of the studies, M. G. Maksud, PhD, in the studies of adults, and Julio Cesar Reina, MD, in the studies on children. The association with these investigators has been particularly rewarding. My gratitude is also expressed to Ms. Therese Gauthier for her careful attention to detail in the original typescript.

REFERENCES

Allen, T.H., Welch, B.E., Trujillo, T.T. & Roberts J.E. (1959). Fat, water and tissue solids of the whole body less its bone mineral. Journal of Applied Physiology, 14, 1009-1012.

Araujo, J., Sanchez, G., Gutierrez, J. & Perez, F. (1970). Cardiomyopathies of obscure origin in Cali, Colombia: clinical, etiologic and laboratory aspects. American Heart Journal, 80, 162-170.

Areskog, N-H., Selinus, R. & Vahlquist, B. (1969). Physical work capacity and nutritional status in Ethiopian male children and young adults. American Journal of Clinical Nutrition, 22, 471-479.

Arteaga, L.A. (1976). The nutritional status of Latin American adults. Basic Life Science, 7, 67-76.

Åstrand, I. (1967). Degree of strain during building work as related to individual aerobic capacity. Ergonomics, 10, 293-303.

Åstrand, P-O. (1952). Experimental Studies of Physical Working Capacity in Relation to Sex and Age. Copenhagen: Munksgaard.

Åstrand, P-O. & Rodahl, K. (1970). Textbook of Work Physiology, p. 292. New York: McGraw-Hill.

Åstrand, P-O. & Rodahl, K. (1977). Textbook of Work Physiology. New York: McGraw-Hill.

Barac-Nieto, M., Spurr, G.B., Maksud, M.G. & Lotero, H. (1978a). Aerobic work capacity in chronically undernourished adult males. Journal of Applied Physiology: Respiratory Environmental Exercise Physiology, 44, 209-215.

Barac-Nieto, M., Spurr, G.B., Lotero, H. & Maksud, M.G. (1978b). Body composition in chronic undernutrition. American Journal of Clinical Nutrition, 31, 23-40.

Barac-Nieto, M., Spurr, G.B., Lotero, H., Maksud, M.G. & Dahners, H.W. (1979). Body composition during nutritional repletion of severely undernourished men. American Journal of Clinical Nutrition, 32, 981-991.

Barac-Nieto, M., Spurr, G.B., Dahners, H.W. & Maksud, M.G. (1980). Aerobic work capacity and endurance during nutritional repletion of severely undernourished men. American Journal of Clinical Nutrition, 33, 2268-2275.

Barac-Nieto, M., Spurr, G.B. & Reina, J.C. (1984). Marginal malnutrition in school-aged Colombian boys: body composition and maximal O_2 consumption. American Journal of Clinical Nutrition, 39, 830-839.

Barac-Nieto, M.   (1984)   Body composition and physical work capacity in undernutrition.   In: P.L. White & N. Selvey (eds.),   Malnutrition: Determinants and Consequences, p. 165-178. New York: A.R. Liss.

Bengoa, J. M. & Denoso, G. (1974).   Prevalence of protein-calorie malnutrition (1963-73). PAG Bulletin, **4**, 24.

Berg, A. (1973).   The Nutrition Factor: its role in National Development. Washington, D.C.: The Brookings Institution.

Berg, A. (1981).   Malnourished People: A Policy View. Washington, D.C.: The World Bank.

Bergstrom, J., Hermansen, L., Hultman, E. & Saltin, B. (1967).   Diet, muscle glycogen and physical performance.   Acta Physiologica Scandinavica, **71**, 140-150.

Brooks, R. M., Latham M. C. & Crompton, D. W. T. (1979).   The relationship of nutrition and health to worker productivity in Kenya.   East African Medical Journal, **56**, 413-421.

Consolazio, C.F., Nelson, R.A., Johnson, H.L., Matoush, L.O., Krzywicki, H.J. & Issac, G.H. (1967).   Metabolic aspects of acute starvation in normal humans: Performance and cardiovascular evaluation. American Journal of Clinical Nutrition, **20**, 684-693.

Cooper, D.M., Weiler-Ravell, D., Whipp, B.J. & Wasserman, K. (1984). Aerobic parameters of exercise as a function of body size during growth in children.   Journal of Applied Physiology: Respiratory Environmental Exercise Physiology, **56**, 628-634.

Correa, P., Restrepo, C., Garcia, C. & Quiroz, A. (1963).   Pathology of heart diseases of undetermined etiology which occur in Cali, Colombia. American Heart Journal, **66**, 534-593.

CVC (1984).   Corporation Autonoma Regional del Cauca.   División Estudios Technicos - Sección de Hidroclimatologiá.   Rio Melendez., Cali, Colombia.

Daniels, J., Oldridge, N., Nagel, F. & White, B. (1978).   Differences and changes in VO_2 among young runners 10 to 18 years of age. Medicine and Science in Sports, **10**, 200-203.

Davies, C.T.M. (1973a).   Physiological responses to exercise in East African children. II. The effects of shistosomiasis, anaemia and malnutrition. Journal of Tropical Pediatrics and Environmental Child Health, **19**, 115-119.

Davies, C.T.M. (1973b).   Relationship of maximum aerobic power output to productivity and absenteeism of East African sugar cane workers. British Journal of Industrial Medicine, **30**, 146-154.

Davies, C.T.M., Brotherhood, J.R., Collins, K.J., Dore, C., Imms, F., Musgrove, J., Weiner, J.S., Amin, M.A., Ismail, H.M., El Karim, M., Omer, A.H.S. & Sukkar, M.Y. (1976).   Energy expenditure and phsyiological performance of Sudanese cane cutters. British Journal of Industrial Medicine, **33**, 181-186.

Donovan, C.M. & Brooks, G.A. (1977).   Muscular efficiency during steady-state exercise.   II.   Effects of walking speed and work rate.   Journal of Applied Physiology: Respiratory Environmental Exercise Physiology, **43**, 431-439.

Durnin, J.V.G.A. (1982).   Muscle in sports and medicine - nutrition and muscular performance. International Journal of Sports Medicine, **3**, 52-57.

Durnin, J.V.G.A. & Passmore, R. (1967). Energy, Work and Leisure. London: Heinemann.

Edmundson W. (1979). Individual variations in basal metabolic rate and mechanical work efficiency in East Java. Ecology of Food and Nutrition, **8**, 189-185.

Edmundson, W. (1980). Adaptation to undernutrition: how much food does man need? Social Science and Medicine, **14D**, 119-126.

Frisancho, A.R. (1974). Triceps skinfold and upper arm muscle size norms for assessment of nutritional status. American Journal of Clinical Nutrition, **27**, 1052-1058.

Gaesser, G.A. & Brooks, G.A. (1975). Muscular efficiency during steady-state exercise: effects of speed and work rate. Journal of Applied Physiology, **38**, 1132-1139.

Gillanders, A.D. (1951). Nutritional heart disease. British Heart Journal, **13**, 177-196.

Gollnick, P.D., Piehl, K., Saubert, C.W., Armstrong, R.B. & Saltin, B. (1972). Diet, exercise and glycogen changes in human muscle fibers. Journal of Applied Physiology, **33**, 421-425.

Halloszy, J.O. (1967). Biochemical adaptations in muscle. Effects of exercise muscle. Journal of Biological Chemistry, **242**, 2278-2282.

Hanson-Smith, F.M., Maksud, M.G. and Van Horn, D.L. (1977). Influence of chronic undernutrition on oxygen consumption of rats during exercise. Growth, **41**, 115-121.

Hansson, J.E. (1965). The relationship between individual characteristics of the worker and output of logging operations. Studia Forestalia Suecia No. 29, pp. 68-77. Stockholm: Skogshogskolan.

Heymsfield, S.B., Stevens, V., Noel, R., McManus, C., Smith, J. & Nixon, D. (1982). Biochemical composition of muscle in normal and semistarved human subjects: relevance to anthropometric measurements. American Journal of Clinical Nutrition, **36**, 131-142.

Immink, M.D.C., Viteri, F.E., Flores, R. & Torún, B. (1984). Microeconomic consequences of energy deficiency in rural populations in developing countries. In: E. Pollit & P. Amante (eds.), Energy Intake and Activity, pp. 355-376. New York: Liss.

Keys, A, Brozek, J., Henschel, A., Mickelsen, O. & Taylor, H.L. (1950). The Biology of Human Starvation. University of Minneapolis Press: Minneapolis.

Lopes, J., Russell, D.M., Whitwell, J. & Jeejeebhoy, K.N. (1982). Skeletal muscle function in malnutrition. American Journal of Clinical Nutrition, **36**, 602-610.

Maksud, M.G., Spurr, G.B., & Barac-Nieto, M. (1976). The aerobic power of several groups of laborers in Colombia and the United States. European Journal of Applied Physiology, **35**, 173-182.

Malina, R.M. (1973). Ethnic and cultural factors in the development of motor abilities and strength in American children. In: G.L. Raric (ed.), Physical Activity, Human Growth and Development, p. 333-364. New York: Academic Press.

Martorell, R., Lechtig, A., Yarbrough, C., Delgado, H., & Klein, R.E. (1978). Small stature in developing nations: its causes and implications. In: S. Margen & R.A. Ogar (eds.), Progress in Human Nutrition, Vol. 2, pp. 142-156. Connecticut: Avi Publishing Company.

Michael, E.D., Hutton, K.E. & Horvath, S.M. (1961). Cardiorespiratory responses during prolonged exercise. Journal of Applied Physiology, **16**, 997-1000.

Montoye, H.J. (1982). Age and oxygen utilization during submaximal treadmill exercise in males. Journal of Gerontology 1982, **37**, 396-402.

Morrison, J.F. & Blake, G.T.W. (1974). Physiological observations on cane cutters. European Journal of Applied Physiology, **33**, 247-254.

Pařízková, J. (1961). Total body fat and skinfolds in children. Metabolism, **10**, 794-807.

Pařízková, J. (1977). Body Fat and Physical Fitness. The Hague: Martinus Nijhoff B.V.

Pate, R.R. (1981). Oxygen cost of walking, running and cycling in boys and men. Medicine and Science in Sport and Exercise 1981, **13**, 123-124(abstr).

Raju, N.V. (1974). Effect of early malnutrition on muscle function and metabolism in rats. Life Science, **15**, 949-960.

Reddy, V. (1981). Protein energy malnutrition: an overview. In: A.H. Harper & G.K. Davis (eds.), Nutrition in health and disease and international development, pp. 237-246. New York: A.R. Liss.

Robinson, S. (1938). Experimental study of physical fitness in relation to age. Arbeitsphysiol, **10**, 251-333.

Rueda-Williamson, R., Luna-Jaspe, H., Ariza, J., Pardo, F. & Mora, J.O. (1969). Estudio seccional de crecimiento, desarrollo y nutricion en 12, 138 ninos de Bogotá, Colombia. Pediatria, **10**, 337-349.

Satyanarayana, K., Nadamuni Naidu, A., Chatterjee, B. & Narasinga Rao, B.S. (1977). Body size and work output. American Journal of Clinical Nutrition, **30**, 322-325.

Satyanarayana, K., Nadamuni Naidu, A. & Narasinga Rao, B.S. (1978). Nutrition, physical work capacity and work output. Indian Journal of Medical Research, **68** (suppl.), 88-93.

Satayanarayana, K., Nadamuni Naidu, A & Narasinga Rao, B.S. (1979). Nutritional deprivation in childhood and the body size, activity and physical work capacity of young boys. American Journal of Clinical Nutrition, **32**, 1769-1775.

Shephard, R.J., Lavallée, H., Larivière, G., Rajic, M., Brisson, G.R., Beucage, C., Jéquier, J-C., & LaBarre, R. (1974). La capacite physique des enfants canadiens: une comparison entre les enfants canadiens-francias, canadiens-anglais et esquimaux: I. Consommation maximale d'oxygene et débit cardiaque. Union Med., **103**, 1767-1777.

Smil, V. (1979). Energy flow in the developing world. American Scientist, **67**, 522-531.

Soedjatmoko, K. (1981). The challenge of world hunger. In: A.E. Harper & G.K. Davis (eds.), Nutrition in Health and Disease and International Development, pp. 1-16. New York: A.R. Liss.

Spurr, G.B. (1983). Nutritional status and physical work capacity. Yearbook of Physical Anthropology, **26**, 1-35.

Spurr, G.B. & Barac-Nieto, M. (1984). Efficiency of submaximal treadmill work in malnourished adult males. Fed. Proc. **43**, 988 (abstr.)

Spurr, G.B., Barac-Nieto, M. and Maksud, M.G. (1975).   Energy expenditure cutting sugar cane. Journal of Applied Physiology, **39**, 990-996.

Spurr, G.B., Maksud, M.G. & Barac-Nieto, M. (1977a).   Energy expenditure, productivity, and physical work capacity of sugar cane loaders. American Journal of Clinical Nutrition, **30**, 1740-1746.

Spurr, G.B., Barac-Nieto, M. & Maksud, M.G. (1977b).   Productivity and maximal oxygen consumption in sugar cane cutters. American Journal of Clinical Nutrition, **30**, 316-321.

Spurr, G.B., Barac-Nieto, M. & Maksud, M.G. (1977c).   Efficiency and daily work effort in sugar cane cutters. British Journal of Industrial Medicine, **34**, 137-141.

Spurr, G.B., Reina, J.C., Barac-Nieto, M. & Maksud, M.G. (1982).   Maximum oxygen consumption of nutritionally normal white, mestizo and black Colombian boys 6-16 years of age. Human Biology, **54**, 553-574.

Spurr, G.B., Reina, J.C., Dahners, H.W. & Barac-Nieto, M. (1983a).   Marginal malnutrition in school-aged Colombian boys: functional consequences in maximum exercise. American Journal of Clinical Nutrition, **37**, 834-847.

Spurr, G.B., Reina, J.C. & Barac-Nieto, M. (1983b).   Marginal malnutrition in school-aged Colombian boys: anthropometry and maturation. American Journal of Clinical Nutrition, **37**, 119-132.

Spurr, G.B., Barac-Nieto, M., Reina, J.C. & Ramirez, R. (1984).   Marginal malnutrition in school-aged Colombian boys: efficiency of treadmill walking in submaximal exercise. American Journal of Clinical Nutrition, **39**, 452-459.

Tanner, J.M. (1961).   Growth at Adolescence. Oxford: Blackwell.

Tanner, J.M. (1976).   Growth as a monitor of nutritional status.   Proceedings of Nutr. Soc. **35**, 315-322.

Tasker, K. & Tulpule, P.G. (1964).   Influence of protein and calorie deficiencies in the rat on the energy-transfer reactions of the striated muscle. Biochem. Journal, **92**, 391-398.   In: R.A. McCance & E.M. Widdowson (eds.),   Calorie Deficiencies and Protein Deficiencies, pp. 289-299. Boston: Little, Brown.

United Nations (1980).   Demographic Year Book (1979).   Department of International Economy and Social Affairs.   New York: United Nations.

Viteri, F.E. (1971).   Considerations on the effect of nutrition on the body composition and physical working capacity of young Guatemalan adults. In: N.S. Scrimshaw & A.M. Altshull (eds.), Amino Acid Fortification of Protein Foods, pp. 350-375. Cambridge, Mass.:   MIT Press.

Viteri, F.E., and Torún, B. (1975).   Ingestión calorica y trabajo físico de obreros agrícolas en Guatemala.   Efecto de la suplementación alimentaria y su lugar en los programs de salud. Bol. Of. Sanit. Panamer., **78**, 58-74.

Viteri, F.E., Torún, B., Immink, M.D.C. & Flores, R. (1981).   Marginal malnutrition and working capacity. In: A.E. Harper & G.K. Davis (eds.), Nutrition in Health and Disease and International Development, pp. 277-283. New York: A.R. Liss.

Wolgemuth, J.C., Latham, M.C., Hall, A., Chester, A. & Crompton, D.W.T. (1982).   Worker productivity and the nutritional status of Kenyan road construction laborers. American Journal of Clinical Nutrition, **36**, 68-78.

Zavaleta, A.N. & Malina, R.M. (1980).   Growth, fatness and leanness in Mexican-American children. American Journal of Clinical Nutrition, **33**, 2008-2020.

MARGINAL ENERGY MALNUTRITION:
SOME SPECULATIONS ON ENERGY SPARING MECHANISMS

A. FERRO-LUZZI

National Institute of Nutrition, Rome, Italy

INTRODUCTION

Energy nutritional status is a dynamic condition that might be described as the extent to which improvements can be made in terms of functions and capacities (such as work capacity and performance, reproductive competence, disease response, social and behavioral functions, growth performance, etc.) by changes in energy balance, its components, or the rate of energy flux (Ferro-Luzzi & Norgan, 1984a).

This definition stresses the need to induce a perturbation to evaluate energy nutritional status. Static measurements such as the size of body energy stores or the level of energy intake, output or balance, are unable to capture the dynamic nature of energy status, introduce an element of circular reasoning and can lead to very misleading results. In fact, except for relatively short periods of acute shortage of food, as in famines, droughts, and other man-made or god-sent adversities, people do basically subsist in a state of long-term energy equilibrium. This equilibrium, however, may be variously achieved. Figure 1 encompasses the known or postulated mechanisms called upon to save, or waste, energy under conditions of energy disbalance. Energy equilibrium type 1 is intrinsically different in quality, nature, impact and consequences from equilibrium conditions types 2 and 3.

The latter condition, namely the type of equilibrium achieved when energy intake is smaller than output, is of enormous importance as it is assumed or suspected that it represents the condition shared by a large proportion of humanity living in less-developed countries (LDC). This debatable assumption is largely derived from the observation that, in these individuals, theoretical energy requirements are often not met and they appear to subsist on intakes barely above their basal metabolic rate (BMR). Other indirect evidence is based upon the measurable consequences of the presumedly enacted survival strategies, namely miniaturisation of body dimensions, decrease of physical activity and

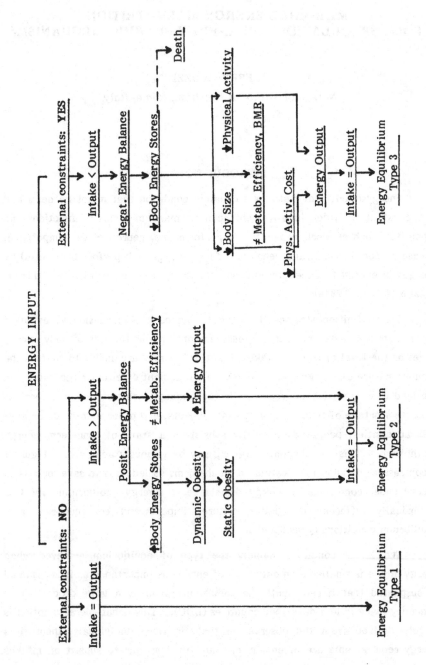

Figure 1: Postulated mechanisms for achieving equilibrium under different energy balance conditions

improved metabolic efficiency of energy utilisation. Thus, growth retardation in children and small attained size in adults, a common feature in LDC populations, are conventionally considered as the tangible expression of one mode of adaptive strategy of individuals and communities facing conditions of restricted food availability. Evidence of recourse to other forms of adaptive strategy, namely the metabolic and behavioral mechanisms, is sparse, inconclusive and controversial.

While much more research is warranted to reach an understanding of the combination of adaptive mechanisms (if any) which is implemented under real life conditions in free-living populations to achieve energy equilibrium, it would nevertheless be highly desirable to have an idea of the nominal size of the energy sparing that could be achieved by these means.

The extent to which any of the three postulated mechanisms, either separately or in combination, could theoretically contribute to restore equilibrium in situations of negative energy balance, have been calculated and are presented in the following paragraphs.

BODY DIMENSION

There is little doubt that the experimental response of a growing animal to restricted food availability, is to reduce its rate of growth. Animal studies are countless, and the evidence is fairly robust.

For what concerns the human being, it has been already mentioned (see also Martorell, this volume) that the short stature in developing countries is conventionally considered to be the outcome of adaptive processes to low dietary intake. The tacit assumption is that the growth potential is identical for all populations, with few notable exceptions. This assumption is based, among others, on the observation that although the height of individuals from low and high socioeconomic conditions can exhibit, within the same ethnic group, a marked difference, the average stature of the high socioeconomic strata is comparable on a cross-cultural basis (Ferro-Luzzi, 1984).

To exemplify this point, the difference in growth performance of a representative cross-sectional sample of about 70,000 urban and rural Italian children (Ferro-Luzzi, 1966) is shown in Figure 2.

The children were measured in the 1950's when several socioeconomic indicators and vital statistics (GNP, housing conditions, food availability, sanitation, etc.) suggested the existence of unfavorable environmental factors capable of interfering with the full expression of the genetic potential for

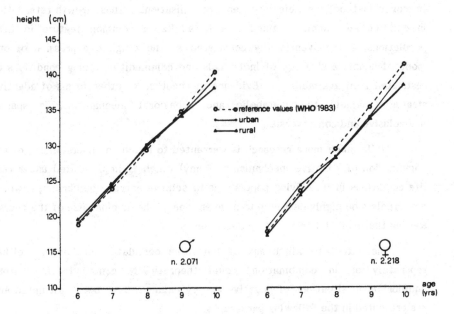

Figure 2:   Height (P₅₀) for a representative sample of urban and rural children
measured in the 1950s.   (Adapted from Ferro-Luzzi, 1966.)

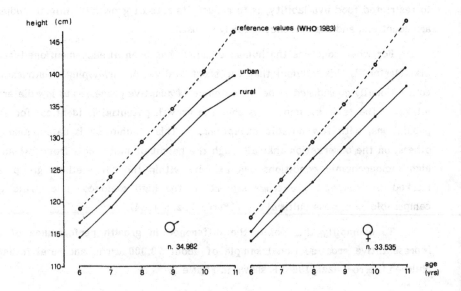

Figure 3:   Mean height of a representative sample of urban and rural children
from Abruzzi, low-income Italian region.   (Adapted from D'Amicis
et al, 1979.)

growth.   The figure illustrates also that the children were shorter than their wealthier counterparts from other developed nations (WHO, 1983).   More recent studies (D'Amicis, Ferrini, Ferro-Luzzi, Maiale & Peruzzi, 1979) have shown that this difference has largely disappeared and that the Italian rural children grow now at the same rate as urban ones and also that both now attain the International Reference Standard (Figure 3).

As far as the LDC are concerned, while small groups of economically privileged healthy children of various ethnic groups in LDC have been reported to have a growth performance similar to that of Western reference standards, the great majority of children show that by the age of 1, or at the latest at 2, they have accumulated a sizeable growth retardation as compared to developed countries, and that there is scarcely any sign of later catch-up growth, the end result being a shorter and lighter adult (Eveleth & Tanner, 1976).

All this suggests the overriding influence of environmental and nutritional stresses and confirms that the divergence of physical dimensions from the norm is not only commonly associated with underdevelopment but also that it can be safely assumed to represent the tangible outcome of an adaptive strategy to nutritional deprivation or, more precisely, to environmental stresses.

Energy saving associated with reduced body dimensions

Smaller body size should, theoretically, decrease the energy cost of physical activities, especially those entailing body displacement.   If, for the sake of simplicity, we assume that this reduction is linearly related to body weight, we can calculate the potential amount of energy spared by means of this adaptive mechanism.   The calculations and the results are shown in Table 1.

For an adult male of standard height (175 cm), appropriate weight (70 kg) and body mass index (BMI 23) we can calculate his energy requirement using the approach adopted by the recent FAO/WHO/UNU report on Energy and Protein Requirements (FAO/WHO/UNU, 1985).   This calculation is based on the predicted BMR and on a factor applied to it describing the intensity of physical activity.

Table 1 illustrates three levels of activity, at a survival minimum (BMR x 1.2), at a low/intermediate level (BMR x 1.4), and at an observed average level (BMR x 1.86) derived from several studies on a total of about 600 subjects measured under real life conditions (see later).

If the same active adult was to have undergone, during his development phase, sufficient nutritional stress to suppress the full expression of his genetic

Table 1:   Effect of body dimensions on energy expenditure:
            comparison between the energy requirement at various
            levels of activity of a well-nourished and of a stunted
            adult man.

	Well nourished		Stunted	
Height, cm	175		165	
Weight, kg	70		60	
BMI	23		22	
BMR, kcal/d	1750		1597	
Activity levels				
at 1.2 BMR, kcal/d	2100	(30)*	1916	(32)
at 1.4 BMR, kcal/d	2450	(35)	2236	(37)
at 1.86 BMR, kcal/d	3255	(47)	2970	(50)

*   kcal/kg

potential for growth, and he had grown to an attained stature of 165 cm, by no means an uncommon value for LDC peoples, the above calculation would give the results shown in the second column of Table 1.   His predicted BMR would be 1597 kcal/d, with a saving of 153 kcal/d as compared to his well-fed, fully-grown counterpart.   If he was required to function in daily life at 1.86 times his BMR, the stunted individual would spend 2970 kcal/d and save 285 kcal/d, equivalent to about one hour of medium intensity physical work.   Still on this line of reasoning, the net daily energy available for activity at BMR x 1.86 would be 1505 kcal/d net for the normal individual, and 1373 for the stunted one.

These hypothetical figures, however, need a closer scrutiny, as it can be seen that the lighter individual appears to spend, for the same activity level, more energy per unit of body mass than the heavier individual.   This is mostly the carry-over effect of the predicted BMR change and of the way of expressing it in terms of mass instead of surface.   This phenomenon has in the past been the subject of much controversy and dispute (Kleiber, 1947; Brozek & Grande, 1955) and it falls outside the scope of this paper to enter into the, as yet unresolved, debate.   We can try, however, to reason on a different basis. If the net 1505 kcal/d (21.5 kcal/d) available to the well-nourished individual for physical activity at a level of BMR x 1.86 were available to the stunted individual, the latter would spend a total of 3102 kcal/d (51.7 kcal/d) and hence function at a level of BMR x 1.94.   However, a more realistic approach might be to standardise the available net energy on the basis of body weight, i.e. 21.5 kcal/kg x 60 kg = 1290 kcal/d.   In this case, the total energy expenditure of the stunted individual would be 2887 kcal/d, or 48.1 kcal/kg; and he would have an activity level of BMR x 1.81.   The estimated energy saving would be, in this case, 368 kcal/d net.

METABOLIC ADAPTATION

Another postulated pathway for energy saving is represented by the reduction of BMR, over and above that accounted for by body weight and fat-free mass losses. A decline of BMR in conditions of prolonged semi-starvation of previously well-fed subjects has been described by several authors (Benedict, 1915; Benedict, Miles, Roth & Smith, 1919), and often reviewed. The reduction that is observed in BMR occurs mostly on the first weeks of exposure to semi-starvation or fasting. The drop in BMR, which is not accountable on the basis of active tissue loss, has been ascribed to a true, hormonal-mediated change in the metabolic efficiency of energy utilisation. There is a consensus that this reduction can reach a maximum of 16% (James & Shetty, 1982).

Regrettably, the evidence that such a mechanism of improved efficiency of energy utilisation is operative in free-living populations with low energy intakes is tenuous and contradictory. The lower BMR consistently reported for Indians (Banerjee & Bhattacharya, 1964; Soares & Shetty, 1984) only adds to the confusion as the study subjects were largely of wealthy background and thus unlikely to have been exposed to low energy intakes. Even more intriguing data come from East Java (Edmunson, 1979), where the BMR of peasant farmers of similar body size ranged from as low as 0.43 kcal/min to as high as 1.84 kcal/min, both quite disconcerting figures, especially when reported without any special comment.

Energy saving associated with BMR reduction

If, nonetheless, we accept the possibility that such forms of adaptation can take place, the calculation of the theoretical saving of energy contributed by this mechanism is shown in Table 2. The calculations assume that the net cost of physical activity expressed in terms of body weight does not change, i.e. that for the same activity, the adapted individual will spare only the amount attributable to the decline in BMR. This is far from being confirmed, and the whole issue of the relationship between body weight and the energy cost of physical activity is very incompletely understood. However, this assumption is the basis of the rationale for the use of published values for the energy cost of various activities in the evaluation of the energy expenditure of other individuals and groups. The FAO/WHO/UNU Expert Committee on energy and protein requirements seems to share this view and has stated that "regardless of body weight, the same multiple of BMR can be used to cost each activity" (FAO/WHO/UNU, 1985).

From Table 2 it can be seen that whereas a 14% reduction of body size causes a 9% reduction in energy output and the 16% decline in BMR causes the

Table 2:   Individual and combined postulated effects of adaptation to low energy intakes.

	Normal	'Adapted'		
		For BMR	For size	For size & BMR
Height, cm	175	175	165	165
Weight, kg	70	70	60	60
BMI	23	23	22	22
BMR, kcal/d	1750	–	1597	–
Adapted BMR, kcal/d	–	1470	–	1341
Expenditure at 1.86 BMR, kcal/d	3255	2734	2970	2494
Net energy saving over "normal", kcal/d	–	521	285	761
Net energy for work, kcal/d	1505	1264	1373	1153

expected 16% decline in total energy expenditure, the combined decrease in body size and in BMR produces a 23% sparing in energy expenditure, no doubt an impressive saving in terms of survival at reduced intakes. Such saving would be obtained while maintaining a physical activity comparable to that of the well-nourished counterpart. The highest reduction appears to be achieved by the combined metabolic and biological mechanisms (761 kcal/d) and the lowest by the biological mechanism only (285 kcal/d). However, if we consider the net energy available for physical work, the best means in terms of energy sparing appears to be the process of body miniaturisation, while reducing both BMR and body size appears to be marginally the least successful (1153 kcal/d).

PHYSICAL ACTIVITY

The next mechanism available to individuals and communities for restoring energy equilibrium under conditions of restricted food availability is represented by adjustments of physical activity. In spite of the fact that this behavioral mechanism might potentially be the most powerful one, direct evidence of reduced physical activity in response to low energy intakes is scarce. A saving of about 1000 kcal/d would theoretically be possible by dropping from an activity level of BMR x 1.86 down to the survival threshold of BMR x 1.2. The latter value, however, is scarcely compatible with active productive life, and I shall not comtemplate here a reduction of this size. In the Minnesota Study (Keys, Brozek, Henschel, Mickelson & Taylor, 1950), the best ever produced on this topic, a drop occurred in the level of physical activity, from BMR x 2.17 to BMR x 1.63. In other words, a fall in gross energy expenditure from 50 kcal/kg to 30 kcal/kg, and in terms of net energy, from 23 kcal/kg to 9 kcal/kg, respectively a 60% and a 40% decline. The subjects, however, were on a fixed exercise routine, and we ignore what would have been their otherwise spontaneous, unregulated response to prolonged semistarvation. Moreover, although this elegant experiment indicates that physical activity can be severely depressed and that this depression can substantially contribute to restore energy equilibrium, the experimental conditions under which this and a few other similar studies were conducted are scarcely comparable to real life conditions.

In real life, the mechanisms that can be called upon to restore energy equilibrium by adjusting energy output, may operate either by cutting out entire blocks of specific tasks, by reducing the time dedicated to various activities, by slowing down the pace, or by diminishing the intensity at which those activities are carried out. To capture the operational mode as well as the extent, the value, and the energy contribution of this mechanism, two types of data are needed; the first is "time-use", i.e. the fractional use of daily time, set within a functionally significant frame (such as the socioeconomic and cultural content of each activity); the second is the energy cost of each activity. Although both are needed to obtain the 24-hr energy expenditure, the first is often omitted in biological papers; on the other hand, papers belonging to the social or economic sphere tend to present time-use data aggregated in ways unsuitable for the present purpose. Scanning the literature, it can be quickly seen that, while far from being as complete as one would wish, the papers published on energy expenditure are more abundant than those on time-use.

Mean daily energy expenditure data are available from a number of studies carried out on samples of adult populations from developing countries. A selection of these papers on the basis of the criteria suggested by Durnin and Ferro-Luzzi (1982) has yielded a total of about 580 surveyed individuals. A summary of these data is reproduced in Tables 3, 4, 5 and 6.

The energy expenditure data have been transformed into activity levels (AL), and are presented as BMR factors. The BMRs have been calculated from the equations proposed in the 1985 FAO/WHO/UNU Report, on the basis of the mean group weight and age. The BMR factors were obtained by dividing the average energy expenditure by the average BMR.

The average AL for LDC adult males is BMR x 1.80, and for adult females, BMR x 1.67. These values can be compared with equivalent data obtained on selected populations from developed countries (DC). An AL of 1.92 was obtained for 215 DC adult males and an AL of 1.45 for 55 DC adult females. These results scarcely suggest any behavioral adaptation to low energy intakes

Table 3: Activity levels (expressed as BMR factors) from selected studies on energy expenditure – Adult females LDC.

Age yrs	Subject No.	Wt. kg	BMR kcal/d	Expenditure kcal/d	Activity level (BMR factor)	Country	Type of Job	Reference
25	10	46.9	1185	1888	1.60	Philippines	Typists	de Guzman et al, 1978
33	14	48.7	1253	2032	1.63	"	Textile – mill workers	de Guzman et al, 1979
34	10	54.0	1299	2032	1.57	"	Housewives	de Guzman et al, 1974
31	12	50.6	1269	2127	1.68	Upper Volta	Farmers, rainy season	Bleiberg et al, 1980
31	12	50.6	1269	2653	2.10	"	Farmers, dry season	Bleiberg et al, 1980
31	14	49.0	1255	1936	1.54	"	Farmers, end harvest	Bleiberg et al, 1981
23	6	49.0	1216	1960	1.62	Guatemala	Peasants	Schutz et al, 1980
28	18	49.2	1219	2008	1.65	"	Peasants, lactating	Schutz et al, 1980
23	23	49.8	1228	1932	1.57	New Guinea	Kaul, non-pregnant, non-lactating	Norgan et al, 1974
40	17	45.9	1228	1692	1.38	"	-do-	Norgan et al, 1974
27	7	51.0	1246	1857	1.49	"	Kaul, pregnant	Norgan et al, 1974
23	31	51.1	1247	2268	1.82	"	Lufa, non-pregnant, non-lactating	Norgan et al, 1974
36	7	46.3	1232	2141	1.74	"	-do-	Norgan et al, 1974
25	7	53.5	1282	2277	1.78	"	Lufa, pregnant	Norgan et al, 1974
25	20	44.5	1150	1984	1.73	Machiguenga	Hunter-gatherers, wet season	Montgomery et al, 1977
27	8	44.3	1147	1960	1.71	"	Hunter-gatherers, dry season	Montgomery et al, 1977
M 28	216	49.0	1231	2050	1.67			

Table 4: Activity levels from selected studies on energy expenditure - Adult females DC.

Age yrs	Subject No.	Wt. kg	BMR kcal/d	Expenditure kcal/d	Activity level (BMR factor)	Country	Type of Job	Reference
19	4	56.3	1324	2103	1.59	Australia	Students, trainees	McNaughton et al, 1970
26	13	63.7	1432	2103	1.47	U.S.A.	Housewives, pregnant, 1st trimester	Blackburn et al, 1976
26	12	70.4	1531	2294	1.50	U.S.A.	Housewives, pregnant, 2nd trimester	Blackburn et al, 1976
26	14	77.0	1628	2366	1.46	U.S.A.	Housewives, pregnant, 3rd trimester	Blackburn et al, 1976
26	12	62.4	1413	1840	1.31	U.S.A.	Housewives, lactating, 1st trimester	Blackburn et al, 1976
M 25	55	67.7	1491	2154	1.45			

Table 5: Activity levels from selected studies on energy expenditure – Adult males LDC.

Age yrs.	Subject No.	Wt. kg	BMR kcal/d	Expenditure kcal/d	Activity level (BMR factor)	Country	Type of Job	Reference
28	10	53.9	1504	2199	1.46	Philippines	Clerk-typists	de Guzman et al, 1978
26	10	56.3	1540	2701	1.76	"	Shoemakers	de Guzman et al, 1974
32	10	54.8	1515	2486	1.64	"	Jepney drivers	de Guzman, Kalaw et al, 1974
25	25	54.8	1517	2486	1.64	"	Textile-mill workers	de Guzman et al, 1979
(25)	33	(56.0)	1536	2390	1.56	"	Fishermen	de Guzman, 1981
(25)	32	54.0	1505	3035	2.02	"	Sugarcane workers	de Guzman, 1981
28	9	51.5	1467	3298	2.25	"	Rice farmers	de Guzman et al, 1974
45	11	56.5	1534	2127	1.39	Upper Volta	Farmers	Bleiberg et al, 1981
36	23	58.5	1558	2414	1.55	"	Farmers, dry season	Brun et al, 1981
36	16	58.5	1558	3466	2.23	"	Farmers, wet season	Brun et al, 1981
18-29	19	57.4	1562	2629	1.69	New Guinea, Kaul	Subsistence horticulture	Norgan et al, 1974
18-25	28	58.3	1570	2557	1.58	" Lufa	"	Norgan et al, 1974
30+	32	55.1	1518	2151	1.42	" Kaul	"	Norgan et al, 1974
30+	15	56.0	1529	2581	1.69	" Lufa	"	Norgan et al, 1974
39	10	56.0	1529	3274	2.15	Iran, Varanin	Agric. workers, spring	Brun et al, 1979
38	14	56.0	1529	3442	2.25	"	Agric. workers, summer	Brun et al, 1979
35	10	59.0	1563	3633	2.33	"	Agric. workers, autumn	Brun et al, 1979
38	12	59.0	1563	2629	1.68	"	Agric. workers, winter	Brun et al, 1979
30	18	60.1	1599	3705	2.32	Guatemala	Agric. labourers	Viteri et al, 1971
30	15	53.9	1504	2892	1.93	Machiguenga	Horticult. hunter-gatherers	Montgomery et al, 1977
25	8	51.8	1472	3322	2.26	"	-do-	Montgomery et al, 1977
M 30	360	56.2	1535	2751	1.80			

Table 6: Activity levels from selected studies on energy expenditure. Adult males DC.

Age yrs.	Subject No.	Wt. kg	BMR kcal/d	Expenditure kcal/d	Activity level (BMR factor)	Country	Type of Job	Reference
36	38	75.6	1756	2855	1.62	Italy	Shipyard, sedentary work	Norgan & Ferro-Luzzi, 1978
39	37	77.9	1783	3137	1.76	Italy	Shipyard, moderate work	Norgan & Ferro-Luzzi, 1978
37	75	76.1	1762	3309	1.88	Italy	Shipyard, heavy work	Norgan & Ferro-Luzzi, 1978
22	6	64.0	1621	2820	1.70	U.K.	Students	Norgan & Durnin, 1980
19	59	63.3	1647	3776	2.29	U.K.	Army recruits	Edholm et al, 1970
M 32	215	72.4	1729	3313	1.92			

Table 7:   Summary of activity levels observed in developing (LDC) and in developed (DC) countries

	No.	Activity Level on normal BMR*	on "adapted" BMR
Adult males			
LDC	360	1.80	2.13
DC	215	1.92	-
Adult Females			
LDC	216	1.67	1.98
DC	55	1.45	-

*   Predicted by weight and age.

by the LDC groups.   A dimensional adjustment, however, might be inferred from their smaller body size as compared to individuals from developed countries. If we were to postulate that, besides being dimensionally adapted to energy malnutrition, the LDC subjects had also made recourse to the metabolic adaptive strategy (i.e. a 16% drop in BMR), their corrected AL would become 2.13 and 1.98 for men and women respectively (Table 7).   These values are higher than those observed in DC.

The conclusion of this exercise would be that there is no evidence of a low/reduced habitual physical activity in individuals supposedly exposed to marginal energy malnutrition. Of course, one could argue that higher activity levels might have been exhibited, or be required or be desirable, had those same LDC subjects had free access to unlimited energy resources. Furthermore, we ignore, although we can strongly suspect, what would be the behavioral choice of people given the option on the one hand to raise their physical activity capacity to higher levels through liberal access to extra food energy, i.e. more food for more work; and on the other hand to diminish their obligatory work load by providing mechanical means that would allow more efficient use of human energies, i.e. more productivity with less physical work and lower energy requirements. No such experiment covering both options has as yet been carried out, to my knowledge; but a few have been made where extra food energy was provided to communities with low energy intakes, and thus assumed to be energy malnourished. Unfortunately only anecdotal or indirect evidence exists on the results in terms of change in energy expenditure.

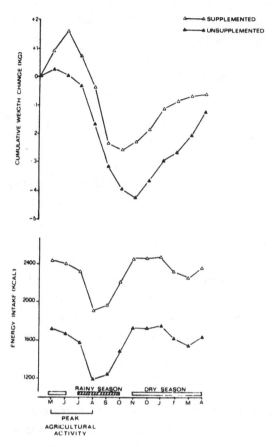

Figure 4: Mean energy intake (kcal/d) and concurrent cumulative body weight changes of supplemented and unsupplemented lactating women in Gambia over 1 year of observation (from Prentice, 1984).

In the Gambia study, conducted on 156 lactating women, a supplement of 723 kcal/d caused a slight and early increase in body weight (1.5 kg), suggesting a preferential diversion of energy to body stores (Prentice, 1984). The remarkable aspect of this study is that the supplement did not prevent the occurrence of seasonal body weight fluctuations that closely paralleled that of the unsupplemented women (Figure 4). However, not all the extra energy provided to these women appeared as body fat, and the authors comment that it is unlikely that the unaccounted supplemental energy (671 kcal/d) could have been entirely used in extra physical activity. They offer, as an explanation, a change in the overall metabolic efficiency in energy handling. If we assume that the women's BMR was depressed by adaptation to low energy intakes prior to the supplementation, then we can theoretically expect that after supplementation their BMR might have increased by 16%, i.e. by about 203 kcal/d, thus leaving 470 kcal/d unaccounted for.   This value represents an increase in net energy expenditure of about 1 kcal/min for about 8 hours a day, over and above that spent prior to supplementation, a more likely value which might easily be missed in a perfunctory observation of patterns of physical activity.   It is regrettable that no actual measurement of BMR and activity levels of these women has been made.

Similar seasonal body weight fluctuations in the absence of any apparent external constraint on energy intake (i.e. unvaried food availability) have been described by Spencer and Heywood (1983) in New Guinea subsistence horticulturalists (Figure 5).   The loss of body weight seemed to reflect patterns of agricultural activity more closely than patterns of food resources, suggesting that maintenance of physical activity may have had a higher priority than mobilisation of body energy reserves, even under conditions of liberal energy supply.   Unfortunately, in this case we miss the hard experimental evidence needed to draw solid and appropriate conclusions on the issue of 'plasticity' of physical activity.

Time-Use

Similar considerations apply to the time-use changes in response to energy malnutrition.   Very few studies have appeared in the literataure on time-use on a daily basis in individuals supposedly exposed to energy malnutrition.   Such studies were conducted for different purposes, such as those to evaluate household budgets and productivity patterns (Francois, Dubois & Yai, 1983), to measure energy expenditure (Norgan, Ferro-Luzzi & Durnin, 1974) to examine social, behavioral patterns, etc.   Unhappily, the data have been collected with different techniques, and the results have been aggregated and elaborated in accordance with the stated purpose of the survey and published in a way that renders them generally unsuitable for the comparative investigation of the influence of energy malnutrition on time use.   This state of affairs is largely the result of methodological problems, and will be discussed later.

Furthermore, physical activity is under cortical control, and thus by definition, largely influenced by volition.   However, the nature of the activities which could be affected in the process of restoring energy equilibrium can be extremely varied.   The strategy which might be consciously or unconsciously enacted by individuals and communities is likely to be heavily culture-dependent.

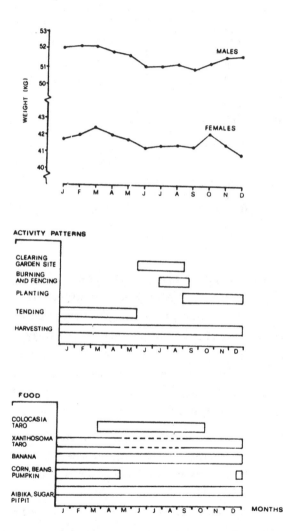

Figure 5: Seasonal changes in weight, agricultural activity patterns and food crops harvesting in Papua New Guinea. (Adapted from Spencer & Heywood, 1982.)

158   *Ferro-Luzzi*

Table 8:   Daily time use by women in four different ecological settings

Activities	Peasants* Ivory Coast mins.	Subsistence Horticulture** New Guinea Highlands mins.	House- wives*** Rural South Italy mins.	Students**** Urban Italy mins.
Sleeping	539	536	519	531
Cooking	131	43	142	4
Domestic tasks	103	86	252	8
Agricultural tasks	58	94	79	–
Conspicuous leisure	40	17	31	79
Income-generating activities	7	17	29	389
Sit/stand/move about	407	478	308	332
Walking	32	141	51	6
Miscellanea	123	28	29	91

*   from Francois, Dubois and Yai (1983)
**   from Norgan, Ferro-Luzzi and Durnin (1974)
***   unpublished data from Ferro-Luzzi et al (1984)
****   unpublished data

Options will be mostly dictated by environmental resources and circumstances, as well as by personal preference and priorities. For all these reasons, it will be extremely difficult, if not impossible, to label any observed 'spontaneous' pattern of activity as the outcome of a behavioral adaptive strategy to low energy intakes. Only the dynamic observation of the departure from a baseline norm following a disturbance might represent a promising approach. Table 8 is presented in order to illustrate how time-motion studies are unable to predict the nature of the existing energy equilibrium.

In Table 8 the results of four studies on time-use of adult women have been summarised. The women were leading completely different lives in different countries. It is conventionally assumed that the two LDC groups might have had a life-long exposure to chronic energy malnutrition while the two other groups would be considered to be well fed. The body sizes ranged from 152 cm and 51 kg, with a BMI of 22 for the New Guinea women (Norgan, Ferro-Luzzi & Durnin, 1974), to 157 cm and 65 kg with a BMI of 27 for the Italian housewives (unpublished results from Ferro-Luzzi, Strazzullo, Scaccini, Siani, Sette, Mariani, Mastranzo, Dougherty, Iacono & Mancini, 1984b). No such data are available for the ivory Coast women (Francois et al, 1983). Time-use data were collected on the New Guinean and on the Italian students by the classical time-and-motion technique (Norgan, Ferro-Luzzi & Durnin, 1974). A similar technique is claimed to have been used for the Ivory Coast group (Francois et al, 1983), although the study and its methodology appears to belong more to the time-budget socioeconomic approach. Besides telling us what is obvious from the direct inspection of the four columns of Table 8, there is no way we can interpret the different patterns of time-use to be an expression of energy malnutrition rather than the outcome of environmental and cultural pressures and habits. As a matter of fact, the only appreciable difference in activity pattern is that of the urban students, the housewives of rural southern Italy being rather similar to their DC counterparts, except for appearing to be overburdened with cooking and with household tasks.

Actual energy expenditure was measured on the New Guinean women and on the Italian students; energy intakes were measured over 42 days for the Italian housewives and have been equated to energy expenditure as the subjects maintained body weights unchanged throughout the period of observation. No measure of energy expenditure was performed on the Ivory Coast women and their energy intakes and body weights have not been published. The energy expenditure (either on an intake or an expenditure basis) was about 34-37 kcal/kg for the two Italian groups (AL = 1.43 to 1.70) and 44 kcal/kg for the New Guineans (AL = 1.80). These values do not support the initial assumption of reduced levels of physical activity as an expression of adaptation to low energy intakes of the New Guinean women.

Methodological considerations

Time-use figures which appear in the literature are generated by a variety of means, and represent a more or less arbitrary aggregation of data. It is most important to discuss this issue in some detail because the technique by which information is gathered, aggregated and analysed has a direct and important bearing on the quality, reliability, precision and interpretability of results of time-use studies. Furthermore, there may be an important carry-over effect on the calculation of energy expenditure when figures of energy cost for single activities are applied to fractions of daily time aggregated on the basis of social and economic considerations.

Time-use data can be collected by means of an observer, unobtrusively marking down a minute-by-minute time-motion diary. This classical technique (Durnin & Passmore, 1967) may be considered as the 'gold standard' against which any other method can and should be tested. Other methods consist usually of more expedient and simplified approaches, based either on interview-recalls covering the immediate past, an 'activity history', etc.

One problem common to these expedient methods is that the recording of the activities is a very delicate and difficult task if the survey is meant to provide the basis for both global energy expenditure and for documentation of the socioeconomic and cultural significance of the various activities.   For example, to calculate the energy spent in any activity, we need to know the postural position in which it is carried out (i.e. sitting, standing, moving about, walking, etc.), the weight of the person and of the implement used, the pace, the environment (such as walking uphill or on uneven ground, etc.), the degree of mechanisation, etc.

The task of the observer, however, must go beyond the annotation of the 'energetic' characteristics of the various activities.   These must be broken down or aggregated into economically, culturally and socially meaningful tasks, if the purpose of the study is to understand how these activities, besides costing time and energy, relate and contribute to the livelihood of the individual and society, the growing of the human capital and the mechanisms by which desirable or undesirable nutritional conditions are generated.   These two sets of information are the basis of a complete understanding of the behavioral adaptation to low energy intakes.

Unfortunately, such extensive data are very difficult to obtain, to record and to analyse.   The simple time-motion diary in its classical and well-known format used by physiologists and nutritionists, is not suitable.   Similarly, the format used in socioeconomic time-budget surveys is seriously deficient in some crucial pieces of information needed to correctly evaluate the energy cost of activities.   A different tool should be designed, compounding these various approaches.   Such a tool does not exist at present and as a consequence surveys designed for a particular purpose are currently being used for another, with highly debateable results (Berio, 1984).   Furthermore, even more expedient approaches (such as activity histories) are currently used without the necessary preliminary validation.   This situation closely parallels the early evolutionary stages of dietary surveys, when dietary recalls and dietary histories were collected and analysed without the necessary validation for reliability, precision, and limits.   Since then, much progress has been made in dietary survey methodology, and we have now a clearer picture of the existence of systematic biases, of the influence of various factors (such as personality of the interviewers, age, sex and cultural background of respondents, etc.) on the dependent variable, i.e. the assessment of dietary intakes.   The highly subjective component of these methods has been recognised to be a source of systematic error, and it has been suggested that appropriate precautions, in terms of validation procedures, need to be taken to improve the reliability and precision of the results.

Nothing like this has as yet been performed for the time-use and energy expenditure surveys, and I submit that this is the most urgent need if we want to identify when and how behavioral adaptation mechanisms to scarce energy resources operate in free living populations throughout the world.   In other words, we want to be able to document and quantify the feeling of improved well-being and inclination to engage more in social activities, reported by the energy supplemented Gambia women (Beaton, 1983), or the games of soccer played after the day's work by Guatemalan agricultural labourers (Viteri & Torun, 1974), following the supply of extra food.

ACKNOWLEDGMENTS

The research work reported here was supported by CNR (National Research Council), Italy, special grant IPRA, subproject 3, Paper No. 409.

REFERENCES

Banerjee, S & Bhattacharya, A.K. (1964).   Basal metabolic rates of boys and young adults of Rajasthan.   Indian Journal of Medical Research,   **52,** 1167-1172.

Beaton, G.H. (1983).   Energy in human nutrition.   Nutrition Today, pp. 6-15.

Benedict, F.G. (1915).   A study of prolonged fasting.   Carnegie Institute, Publication no. 203, Washington DC, p. 416.

Benedict, F.G., Miles, W.R., Roth, P. & Smith, H.M. (1919).   Human vitality and efficiency under prolonged restricted diet.   Carnegie Institute, Publication no. 280, Washington DC, p. 701.

Berio, A.J. (1984).   The analysis of time allocation and activity, patterns in nutrition and rural development planning. Food and Nutrition Bulletin, **6,** 53-68.

Blackburn, M.W. & Calloway, D.H. (1976).   Energy expenditure and consumption of mature, pregnant and lactating women.   Journal of The American Dietetic Association, **69,** 29-37.

Bleiberg, F.M., Brun, T.A. & Goihman, S. (1980).   Duration of activities and energy expenditure of female farmers in dry and rainy seasons in Upper Volta. British Journal of Nutrition, **43,** 71-82.

Bleiberg, F.M., Brun, T.A., Goihman, S. & Lippman, D. (1981).   Food intake and energy expenditure of male and female farmers from Upper-Volta. British Journal of Nutrition, **45,** 505-515.

Brozek, J. & Grande, F. (1955).   Body composition and basal metabolism in man; correlation analysis versus physiological approach.   Human Biology, **27,** 22-31.

Brun, T.A., Geissler, C.A., Mirbagheri, I., Hormozdiary, H., Bastani, J. & Hedayat, H. (1979).   The energy expenditure of Iranian agricultural workers. The American Journal of Clinical Nutrition, **32,** 2154-2161.

Brun, T.A., Bleiberg, F. & Goihman, S. (1981).   Energy expenditure of male farmers in dry and rainy seasons in Upper-Volta.   British Journal of Nutrition, **45**, 67-75.

D'Amicis, A., Ferrini, A.M., Ferro-Luzzi, A., Maiale, G. & Peruzzi, G. (1979). Accrescimento e stato di nutrizione di bambini abruzzesi di 6-10 anni nelle diverse situazioni socioambientali.   SINU, Proceedings 12th Congress, L'Aquila, Italy, 1979.

Durnin, J.V.G.A. & Ferro-Luzzi, A. (1982).   Conducting and reporting studies on human energy intake and output: suggested standards.   The American Journal of Clinical Nutrition, **35**, 624-626.

Durnin, J.V.G.A. & Passmore, R. (1967).   Energy, Work and Leisure.   London: Heinemann Educational Books.

Edholm, O.G., Adam, J.M., Heavy, M.J.R., Wolff, H.S., Goldsmith, R. & Best, T.W. (1970).   Food intake and energy expenditure of army recruits. British Journal of Nutrition, **24**, 1091-1107.

Edmunson, W. (1979).   Individual variations in basal metabolic rate and mechanical work efficiency in East Java.   Ecology of Food and Nutrition, **8**, 189-195.

Eveleth, P.B. & Tanner, J.M. (1976).   Worldwide Variation in Human Growth. London: Cambridge University Press.

FAO/WHO/UNU (1985).   Energy and protein requirements.   Technical Report Series,   Geneva: WHO.

Ferro-Luzzi, A. (1984).   Environment and physical growth.   In: C. Susanne (ed.), Genetic and Environmental Factors during the Growth Period.   New York: Plenum.

Ferro-Luzzi, A. & Norgan, N.G. (1984).   The assessment of energy nutritional status: what is it that is being assessed?   Group of European Nutritionists Workshop: Nutritional status assessment of individuals and population. Gargnano, Italy, 24-26 September.

Ferro-Luzzi, A., Strazzullo, P., Scaccini, C., Siani, A., Sette, S., Mariani, M.A., Mastranzo, P., Dougherty, R.M., Iacono, J.M. & Mancini, M. (1984). Changing the Mediterranean diet: effects on blood lipids.   The American Journal of Clinical Nutrition, **40**, 1027-1037.

Ferro-Luzzi, G. (1966).   The nutritional status of the rural population in 16 Italian communities. Quaderni della Nutrizione, **26**, 94-106.

Francois, P., Dubois, J.L. & Yai, E. (1983).   Depense energetique et budget-temps: an indice d'activite physique fonde sur des donnees collectees en Cote d'Ivoire.   Alimentation et Nutrition, **9**, 31-39.

de Guzman, P.E. (1981).   Energy allowances for the Phillippine population. Proc.   Workshop "Energy expenditure under field conditions", Prague, Czechoslovakia.

de Guzman   P.E., Cabrera, J.P., Basconcillo, R.O., Gaurano, A.L., Yuchingtat, G.P., Tan, R.M., Kalaw, J.M. & Recto, R.C. (1978).   A study of the energy expenditure, dietary intake and pattern of daily activity among various occupational groups. V: Clerk-Typist. Philippine Journal of Nutrition, **31**, 147-156.

de Guzman, P.E., Dominguez, S.R., Kalaw, J.M., Basconcillo, R.O. & Santos, V.F. (1974). A study of the energy expenditure, dietary intake, and pattern of daily activity among various occupational groups.   I. Laguna rice farmers. The Philippine Journal of Science, **103**, 53-65.

de Guzman, P.E., Dominguez, S.R., Kalaw, J.M., Buning, M.N., Basconcillo, R.O. & Santos, V.F. (1974). A study of the energy expenditure, dietary intake and pattern of daily activity among various occupational groups. II. Marikina shoemakers and housewives. Philippine Journal of Nutrition, **27**, 21-20.

de Guzman, P.E., Kalaw, J.M., Tan, R.H., Recto, R.C., Basconcillo, R.O., Ferrer, V.T., Tombokon, M.S., Yuchingtat, G.P. & Gaurano, A.L. (1974). A study of the energy expenditure, dietary intake and pattern of daily activity among various occupational groups. III. Urban Jeepney Drivers. Philippine Journal of Nutrition, **27**, 182-188.

de Guzman, P.E., Recto, R.C., Cabrera, J.P., Basconcillo, R.O., Gaurano, A.L., Yuchingtat, G.P., Abanto, Z.U. & Math, B.S. (1979). A study of the energy expenditure, dietary intake and pattern of daily activity among various occupational groups. VI. Textile-mill workers. Philippine Journal of Nutrition, **32**, 134-148.

James, P.W.T. & Shetty, P.S. (1982). Metabolic adaptations and energy requirements in developing countries. Human Nutrition: Clinical Nutrition, **36C**, 331-336.

Keys, A., Brozek, J., Henschel, A., Mickelson, O. & Taylor, H.L. (1950). The Biology of Human Starvation. Minneapolis, Minn.: University of Minnesota Press.

Kleiber, M. (1947). Body size and metabolic rate. Physiological Reviews, **27**, 511-541.

McNaughton, J.W. & Cahn, A.J. (1970). A study of the energy expenditure and food intake of five boys and four girls. British Journal of Nutrition, **24**, 345-355.

Montgomery, E. & Johnson, A. (1977). Machiguenga energy expenditure. Ecology of Food and Nutrition, **6**, 97-105.

Norgan, N.G. & Durnin, J.V.G.A. (1980). The effects of six weeks of overfeeding on the body weight, body composition and energy metabolism of young men. The American Journal of Nutrition, **33**, 978-988.

Norgan, N.G. & Ferro-Luzzi, A. (1978). Nutrition, physical activity and physical fitness in contrasting environments. In: J. Parizkova & V.A. Rogozkin (eds.), Nutrition, Physical Fitness, and Health: International Series on Sport Sciences, **7**, 167-193. Baltimore: University Park Press.

Norgan, N.G., Ferro-Luzzi, A. & Durnin, J.V.G.A. (1974). The energy and nutrient intake and the energy expenditure of 204 New Guinean adults. Philosophical Transactions of the Royal Society, London, B, **268**, 309-348.

Prentice, A.M. (1984). Adaptations to long-term low energy intake. In: E. Pollit & P. Amante (eds.), Energy Intake and Activity, pp. 3-31. New York: Liss.

Schutz, Y., Lechtig, A. & Bradfield, R.B. (1980). Energy expenditure and food intakes of lactating women in Guatemala. The American Journal of Nutrition, **33**, 892-902.

Soares, M.J. & Shetty, P.S. (1984). Resting metabolic rates of Indian subjects with low bodymass index on normal energy intakes. Euronut. Report No. 5.

Spencer, T. & Heywood, P. (1983). Seasonality, subsistence, agriculture and nutrition in a lowland community of Papua New Guinea. Ecology and Food Nutrition, **13**, 221-229.

Viteri, F.E., Torun, B., Garcia, J.C. & Herrera, E. (1971). Determining energy costs of agricultural activities by respirometer and energy balance techniques. The American Journal of Clinical Nutrition, **24**, 1418-1430.

Protein Foods, pp. 350-375.   Boston: Massachusetts Institute of Technology.

World Health Organisation (1983).   Measuring Change in Nutritional Status. Geneva: World Health Organisation.

GROWTH, STATURE AND MUSCULAR EFFICIENCY

GROWTH, STATURE AND FITNESS OF CHILDREN IN TROPICAL AREAS

J. GHESQUIERE and C. D'HULST

Katholieke Universiteit Leuven, Instituut voor Lichamelijke Opleiding, Leuven, Belgium

INTRODUCTION

Human variability (at adulthood as well as during growth) is greater in the tropics than in any other part of the world. Male adult sizes of almost 185 cm (the Nuer of South Sudan, according to Twiesselman, 1965) or below 145 cm (the Efe and Bambuti of the Ituri forest: 144 cm, according to Gusinde, 1948) have been reported. Such large differences have to be ascribed clearly as much to genetic as to environmental factors, and should show up during growth: genetic, because it is difficult to attribute differences of more than 30% in adult stature exclusively to environmental factors thwarting the genetic potential; environmental, certainly, because it is clear that in this vast region the interplay of environmental factors influence the physical development towards adult size as in no other part of the world.

Genetic factors will have to do with long-range adaptation to the environment. Hiernaux (1977) explains the small size of the pygmy as a gradual adaptation to the heat and moisture of the rain forest: taking a north to south gradient of increasing humidity, he ascribes the difference in size between the Sara of South Tchad, the Tomba and the Twa of the equatorial region in Zaire, in terms of a genetic answer to the problem of thermoregulation. Broadly speaking, this theory holds. However, following a west-to-east gradient along the equator, one encounters decreasing humidity and temperatures, going from the Lake Tumba region of the Twa, whose adult males stand 159 cm (Ghesquiere & Andersen, 1970) to the Ituri-forest of the Bambuti, who stand only 144 cm (Gusinde, 1944; confirmed later). So one must suspect influences other than thermoregulatory adaption.

Clearly the environmental factors at work in the more developed countries of the northern hemisphere are at work in the tropics, even more so.

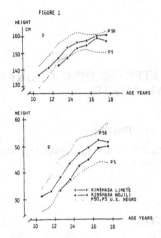

Figure 1:   Stature and weight of Kinshasa schoolgirls from the upper middle-
class district of Limete and a poorer district of Ndjili (Ghesquiere,
1971).   (P5 and P50 are the 5th and 50th percentiles for American
negro schoolchildren. Malina, 1974.)

Differences in nutrition, socioeconomic status of the parents, and between city
dwellers and those of rural areas are far more pronounced here;   one can easily
illustrate how they affect the physical development of the children.

Figure 1 compares schoolgirls from Limete to schoolgirls from Ndjili, a
poorer district, both from the city of Kinshasa, capital of Zaire.   Average
family income at that time (mid-seventies) was 38 Zr (76 $) per month for
Limete and 28 Zr (56 $) per month for Ndjili (Ghesquiere, 1971).   Clearly, the
socioeconomic status of the parents, which directly affects nutrition, hygiene
and access to medication, influences the development of the children.

Living area - city versus countryside - also affects their development as
illustrated in Figure 2, where children from Kinshasa are compared to those of a
large mission school at Lemfu (Lower Zaire) where diet was excellent.   Even in
more rural areas, as in the Ituri district of north-east Zaire, this difference can
be noted on Table 1, comparing the stature of children in Kimodja, a small
village in that area, to those of Bunia, the district capital, both predominantly of
the Bira tribe, and the U.S. negro according to Malina and Roche (1974).

Nutrition in the Ituri forest is not bad, people live in small communities,
and garden culture is adequate if not abundant most of the year.   Still, general
hygiene and access to medication is better in town.   These "adverse conditions"
seem to affect girls less than boys, as already pointed out by Greulich (1951) and
by Tobias (1971), and leads to a greater "crossing over" during growth, and
smaller dimorphism in adult stature.

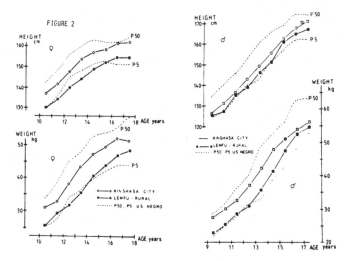

Figure 2:   Stature and weight of girls and boys from urban Kinshasa versus
rural Lemfu (Ghesquiere, 1971); 5th and 50th percentiles for
American negro (Malina, 1974) are shown in dotted lines.

Table 1:   Mean stature

Age (years)	Kimodja Village Boys (cm)	Kimodja Village Girls (cm)	Bunia Town Boys (cm)	Bunia Town Girls (cm)	U.S. Negro Boys (cm)	U.S. Negro Girls (cm)
6.5	111	111	113	115	121	121
10.5	124	125	130	129	143	144
14.5	140	148	153	153	163	161

Comparing height and weight of those African children to American
negroes, the first impression is one of growth retardation.   This we have come
to associate with malnutrition.   However, if we use a simple robusticity index,
putting weight on height as in Figure 3, we see surprisingly little difference
between Africans, Europeans and North Americans, both negro and white.   The
difference in physical growth and development is merely one of size, not so much
of robustness - African boys and girls are smaller.   This is of importance and
will be relevant when we look at motor and physical performance.   If we
compare physical performance, children from tropical areas seem to lag somewhat behind their European and North American counterparts of the same age.
But is this comparison on the basis of age only a fair one?   After all, we see that
most children from tropical areas are physically small, compared to Europeans or
North Americans of the same age.   Would it not be better to compare their
physical   performance   on   the   basis   of   their   physical   growth   and

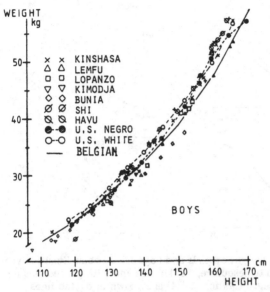

Figure 3:   Stature to weight for African boys compared to Europeans and
North-Americans (Kinshasa according to Twiesselman, 1957; Lemfu:
Ghesquiere, 1971; Lopanzo, Kimodja, Bunia: Nkiama, 1985; Shi,
Havu: collected by Ghesquiere; U.S. white and negro: Malina,
1974; Belgian: Twiesselman, 1969.)

development, as it is obvious that taller boys and girls will have at least some
advantages over smaller boys and girls of the same age?   As we know from
watching sports, a tiny individual hardly has a chance in many events.

Can differences in physical performance in men varying in size,
including children, be explained by dimensional factors?   This has been of
interest to scientists since Gallileo (1638) and Borelli (1685) or before, and
analysed in detail more recently by Adolph (1949), Hill (1950), Asmussen (1955)
and others.   If this fails, then we have to look for an answer in biological
adaptation, or in maturation, or in training or psychological factors.   If we
accept that boys - or girls - retain roughly the same proportions as well as body
composition between, say, the age of 7 and 17 years, but only differ in size, then
the relationship between their surface and volume can easily be derived from the
scale of their stature: their linear dimension, including levers, ranges of joint
motions and muscular contractions will be proportioned, or $L : 1$;   their surface
areas, including cross section area of muscles and bones, hence muscle strength
will be related as $L^2 : 1$;   and their volumes including lung, blood and heart
volumes and weights as $L^3 : 1$.   In accordance with relative size, we would
predict that force applied to an external load should strongly increase with size

(directly related to cross section of the muscle fibre (L^2) and the distance this force is applied (L), hence $L \propto L^3$). Ability at lifting one's own body as in "bent arm hang", "chin-ups", "sit-ups", and "leg fits" should diminish with increasing size (= force times muscle lever must be divided by mass times weight lever = $\frac{L^2 \times L}{L^3 \times L} \propto L^{-1}$. Broad jump or high jump is a question of maximal muscular force over the distance the muscle can shorten before the body-mass leaves the ground, hence proportional to $L^2 \times L/L^3 = 1$; however, the larger boy has the advantage since his centre of gravity is at a higher level before taking off. The time scale is proportional to the length scale; the speed in running is determined by the stride and the number of movements per time unit ($H \propto \frac{1}{L}$). Thus, the maximal speed is proportional to $L/L = 1$, i.e. it should be the same (Åstrand & Rodahl, 1970).

Asmussen (1965) shows that predictions based on these assumptions have to be corrected for qualifying changes during growth; at the age of 11 there are only small differences between shorter and taller boys, while at 14 coordination has improved further, but the taller boys can run faster: they have sexually matured and sex hormones influence their muscular strength. By the age of 18, all boys have matured, and the difference in running speed between boys of different size has vanished. Hence, not only better coordination in both boys and girls may improve their performance, but also the onset of sexual maturation in boys. Although this onset is poorly documented for boys of tropical areas, we have some figures for girls, indicating that this onset may vary a great deal from one ethnic group to the other (Petit-Marie-Heintz, 1963). The same no doubt is true for boys, and will complicate comparison of performance results of children from different areas. Also, the assumption of geometric similarity between children of different ages between 7 and 17 is not entirely correct, and further, children from some tropical areas differ in proportions from Europeans, as we can see from Figure 4 (based on data from Twiesselman (1957, 1969) and data collected by Nkiama (1982-84)). Limbs, especially arms, are much longer in Africans, and may favour performance in jumping and throwing.

Last, the assumption of equal body composition is not correct. In many tasks and fitness tests, the subject is asked to lift or carry his own body. A smaller proportion of "dead weight", or percent body fat, will be an advantage. Judging from skinfold thickness, tropical people have much less fat than Europeans and North Americans. Body fat, calculated from skinfold measurements, was only 8% for a sample of Kivu-Twa (Ghesquiere & Karvonen, 1981) or 6% for Massai! How fit are the children in the tropics? Differences in body size and relative lean body mass, i.e. body composition; differences in degree of

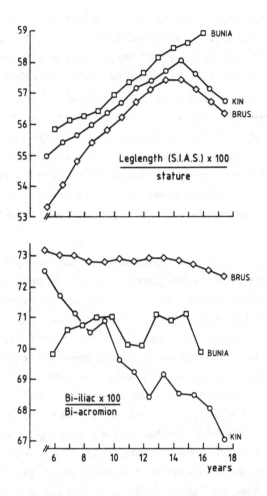

Figure 4:   Body proportions of boys from Bunia (1), Kinshasa (2) and
Brussels (3) (on basis of data collected by Nkiama, 1979 (1), and
Twiesselmann, 1957 (2) and 1969 (3)).

maturation for a given chronological age;   differences in limb to trunk
proportions; - all this on a background of inadequate if not deficient nutritional
conditions - make comparisons difficult.   Furthermore, the data available report
on tests and measurements of a wide variety, with little standardisation in
testing procedure or reporting of results.

In a report to the pre-Olympic Scientific Symposium in Eugene, Oregon,
Malina (1984) gave an extensive review of motor development and performance
of children and youth in undernourished populations of whom, as it happened, the
majority were living in the tropics.   From this review and other isolated reports,

the implications of chronic PEM (protein energy malnutrition) together with other adverse conditions, appear to lead first of all to a high risk of infant- and pre-school mortality. Once the handicaps of this first period are overcome, the child often is able to cope rather well, be it with a stunted physical growth, at par with a reduced muscle mass. Spurr (1984) and Malina (1984) also indicate a delayed motor development, reduced levels of physical activity, and reduced working capacity and efficiency.

Jeliffe and Jeliffe (1969), Viteri (1971), Spurr, Baracnieto and Maksud (1978) and Satayanarayana, Nadidu and Rao (1979) all agree that children, reared under marginal nutritional circumstances, are generally shorter and lighter than better-nourished individuals of the same ethnic group, and show a reduced lean body mass. This should result in a reduction in absolute strength, and as a consequence influence such motor tasks as running, jumping, throwing, involving both motor coordination and strength.

In a study on 350 Zapotec children in south Mexico, 6 to 15 years old, Malina and Buschang (1984) reported that the performance of the boys in the dash and long jump was in proportion to smaller body size in the younger boys, but below that expected for body size at older ages. In contrast, ball-throw for long distances was greater. Comparing these results on Zatopec children to the performance of Papua New Guinean children of similar age, Malina, Shoup and Little (1984) found considerably better running and jumping performances among the latter, whose stature and weight, and hand-grip strength were similar to those of the Zatopec children. Brandt and Broekhoff (1981) compared a group of youngsters, fourth to sixth grade (on average 10 to 12 years old) of coastal highland, and urban Morobo boys of New Guinea to American standards. The Papuan boys performed poorly in shuttle run, pull-ups, sit-ups and distance throw, but performed better in long jump and the 600-yard run. The authors do not consider body size in their analysis, although on average the Papuan boys were shorter and lighter than for U.S. reference figures. According to a Burma Medical Research Council Report (1968) Burmese boys and girls, aged 15 to 17 years, performed poorly compared to North American norms. However, the Papuan and Burmese children jump and run consistently better than would be anticipated from their small body size. In his overview of physical working capacity in undernourished populations - of which again many live in the tropics - Spurr (1983) concludes that malnourished children have lower oxygen uptakes compared to well-nourished children of the same age, especially European and North American. They show a stunted (or, maybe better, reduced) growth, and lower percent of body fat. As a result, however, the maximal oxygen uptake expressed in ml per kilogram body weight does not differ between marginally malnourished and well-nourished boys.   Oxygen uptake measured during leg exercise should be related to the muscle mass available for that exercise (Cotes & Davies, 1969).   If standardised for leg muscle mass, the oxygen uptake is identical in young Africans and Europeans. Spurr (1983) concludes, "there does not appear to be any basic deficit in muscle function in marginally malnourished children, only in the quantity of muscle available for maximal work".

Fitness of youngsters in tropical areas seems almost invariably related to poor nutrition: when one reviews the literature dealing with strength, motor development and performance of children in the tropics, it seems an accepted fact that - with the exception of the few better-off who belong either to the local elite or to expatriate families - nearly all children living in the tropics are considered to be either undernourished, chronically undernourished, mild-to-moderately undernourished, or nutritionally at risk.   The fact, however, is that in large parts of Africa or other tropical areas, nutrition, although not abundant, is adequate.   As Malina (1984) points out, nutritional status is a variable concept, and needs to be more precisely defined;   it is difficult to apply observations from one population to another and further, it should be emphasised that performance of motor tasks is influenced by motivation, competition and cultural differences.

If there seems to be a general consensus that malnutrition will reduce body development and size, it is less clear to what extent this smaller size will affect physical or motor fitness, especially in the ill-defined category of mild-to-moderate PEM.

We often lack precise data on energy intake and expenditure. To some, poor or malnutrition not only reduces size, hence absolute performance, but also performance related to the reduced body size.   Others report no impaired physical performance and functional efficiency (Frinsancho, Velasquez & Sanchez, 1975).   To others still, reduced body size has been associated with superior functional efficiency (Stini, 1975).   Our own observations seem to confirm this hypothesis.   Comparing the motor fitness of three groups of schoolchildren living in Kinshasa, we (Ghesquiere & Eeckels, 1984) found that boys of the shanty town were much reduced in size and absolute strength compared to their better-off compatriots (whose average monthly expenditure per family was more than four times greater) and expatriate boys both attending a private school in the city.   However, when their physical performance was expressed in relation to body size, the shanty boys performed better in most tests (see Figure 5, a, b, c).

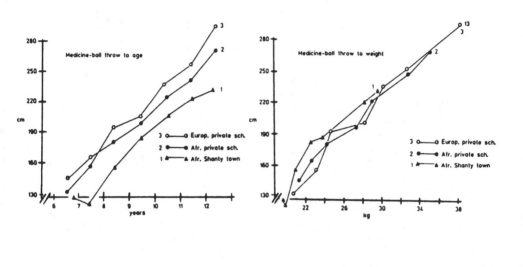

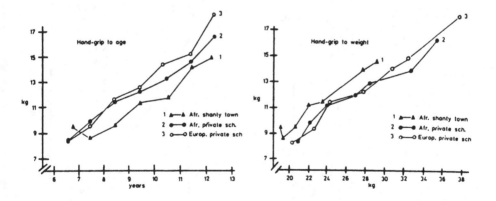

Figure 5:   Anthropometric measurement and motor performance - results of school boys from the upper class African (2) and European (3) private school in Kinshasa versus the lower African Shanty town boys (1) (Ghesquiere & Eeckels, 1984).

5a: Anthropometric measurements

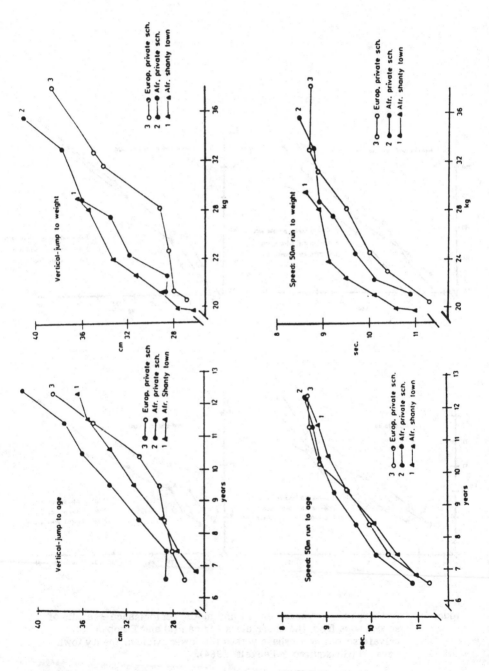

Figure 5b: Medicine ball throw and handgrip.

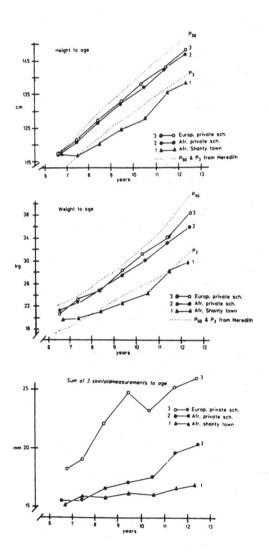

Figure 5c: Vertical jump and 50m run.

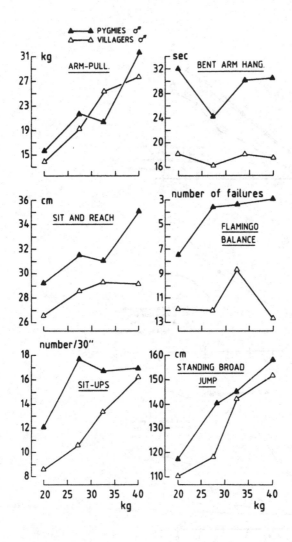

Figure 6:   Relationship between motor performance and weight for Efe
(pygmy) versus Lese village boys (based on observations by
Ghesquiere, Coremans & Nkiama, 1984).

For at least some of the people living in the tropics, motor skills and fitness may have a survival value. Figure 6 compares the performance in such simple motor tasks as bent arm hang, flamingo balance, and sit-ups between children of Efe-pygmies and Lese - villagers sharing the same habitat in the Ituri forest, N.E. Zaïre. The pygmy children performed almost invariably better than the villagers; and their scores were good compared to standards for Belgian or Dutch children of the same body size.

The problem, as Beaton (1983) points out, is that "we have come to accept the notion that the ultimate measure of the adequacy of energy intake is an anthropometric index". This could be misleading. Do we have any evidence that body size and composition is an intermediary variable between food intake and body function? Sukhatme and Margen (1982) have introduced the idea that the energy requirement of an individual varies over a wide range across time and that he possesses the inherent ability to adjust metabolic expenditure of energy. Without joining the controversy over their theory, we may ask ourselves if nature is not proceeding in a similar way when fashioning the human creature. When plenty of food - protein and energy - is available, nature fashions a rather tall, well-functioning creature, endowed with more than necessary (energy) reserves. When less protein and energy foodstuff is available, nature proceeds to build a smaller, but still perfectly functioning creature - with less energy reserves. We should remember that, for people living under tropical conditions of high temperature and humidity, size reduction not only relaxes the pressure on food available, but also on the thermoregulative capacity of the body. Taking a more evolutionary viewpoint, we may ask ourselves if this condition of "nutrition at risk" is not the living condition for which nature has prepared the human race: setting a high genetic potential, knowing by experience that the environment will thwart some of this potential. The result of this shows up clearly in the fitness, motor performance and strength of many of the so-called "growth retarded" or "growth stunted" populations who, in relation to their body size, are just as fit and efficient as their taller counterparts.

REFERENCES

Adolph, E.F. (1974). Quantitative relation in the physiological constitution of mammals. Science 109, 579.

Asmussen, E. (1965). The biological basis of sport. Ergonomics, 8(2), 137-142.

Asmussen, E. & Heeboll-Nielsen, K. (1955). A dimensional analysis of physical performance and growth in boys. Journal of Applied Physiology, 7, 593-603.

Astrand, P.O. & Rodahl, K. (1970). Textbook of Work Physiology, pp. 319-339. New York: McGraw-Hill.

178   *Ghesquiere & D'Hulst*

Beaton, G.H. (1983).   Energy in human nutrition: perspectives and problems.
    Nutrition Reviews, **41**, 325-340.

Brandt, T.R. & Broekhoff, J. (1981).   Physical fitness and motor skills of coastal,
    mountain and urban Morobe boys of Papua, New Guinea,   In: J.B.
    Broekhoff & J. Borms (eds.),   IPCHER Research Symposium, Manila,
    Philippines, pp. 119-131.

Cotes, J.E. & Davies, C.T.M. (1969).   Factors underlying the capacity for
    exercise: a study in physiological anthropometry.   Proceedings of the
    Royal Society of Medicine, **62**, 620-624.

Davies, C.T.M. (1973).   The contribution of leg volume to maximum aerobic
    power output: the effects of anemia, malnutrition and physical activity.
    Proceedings of the Physiological Society, pp. 108-109.

Davies, C.T.M., Barnes, C., Fox, R.H., Ola-Ojikutu, R. & Samueloff, A.S. (1972).
    Ethnic differences in physical working capacity.   Journal of Applied
    Physiology, **33**, 726-732.

Frisancho, A.R., Velasquez, T. & Sanchez, J. (1975).   Possible adaptive
    significance of small body size in the attainment of aerobic capacity
    among high-altitude Quechua natives. In: E.S. Watts, F.E. Johnston &
    G.W. Lasker (eds.),   Biosocial Interrelations in Population Adaptation.
    The Hague; Mouton.

Ghesquiere, J. (1971).   Physical development and working capacity of Congolese,
    In: D.J.M. Vorster (ed.), Human Biology of Environmental Change, pp.
    117-119. London.

Ghesquiere, J. & Andersen, K. (1973).   Similarities and differences in physio-
    logical responses to muscular exercise among primitive African men of
    two genetically different population groups,   In: Selinger (ed.), Physical
    Fitness, pp. 164-176. Prague: Charles University.

Ghesquiere, J. & Eeckels, R. (1984).   Health, physical development and fitness
    of primary chool children in Kinshasa,   In: J. Ilmarinen & I. Valimaki
    (eds.), Children and Sport, pp. 18-30. Berlin/Heidelberg: Springer.

Greulich, W.W., C.S. Crismos & Turner, M.L. (1959).   The physical growth and
    development of children who survived the atomic bombing of Hiroshima
    and Nagasaki. Journal of Pediatrics, **43**, 121-145.

Gusinde, M. (1948). Urwaldmenschen am Ituri. Wien: Springer.

Jeliffe, E.F.P. & Jeliffe, D.B. (1979).   The arm circumference as a public health
    index of protein-caloric malnutrition of early childhood.   Journal of
    Tropical Pediatrics, **15**, 177-260.

Hiernaux, J. (1977).   Long-term biological effects of human migration from the
    African Savanna to the equatorial forest: a case study of human
    adaptation to a hot and wet climate,   In: G.A. Harrison (ed.), Population
    Structure and Human Variation, pp. 187-217.   Cambridge: Cambridge
    University Press.

Malina, R.M. (1984).   Physical activity and motor development performance in
    populations nutritionally at risk,   In: E. Politt & P. Anante (eds.), Energy
    Intake and Activity. New York: Liss.

Malina, R.M. & Buschang, P.H. (1984).   Growth Strength and Motor Performance
    of Zatopec Children. Mexico: Oaxaco.

Malina, R.M., Hamill, P.V.V. & Lemeshow, S. (1974).   Body dimensions and
    proportions, white and negro children 6-11 years.   Washington: Govern-
    ment Printing Office.

Malina, R.M., Shoup, R.F. & Little, B.B. (1985).   Growth Strength and Motor Performance of Manus Children, Papua New Guinea, s.l.

Meredith, H.V. & Nelson, W.E. (1964).   Textbook of Pediatrics.   Philadelphia: Saunders.

Nkiama, E. (1985).   Croissance et Developpement Physique des Enfants Scolarises de Bunia (Zaire).   Master's thesis (unpubl.), Leuven.

Petit-Marie-Heintz, N. (1963).   Croissance et puberte feminine au Rwanda. Memoire academie royale des sciences d'outre-Mer, Science Naturelle et Medicinale, Bruxelles.

Satyanarayana, M.B., Naidu, A.N. & Rao, N. (1979).   Nutritional deprivation in childhood and the body size, activity and physical work capacity of young boys.   American Journal of Clinical Nutrition, **32**, 1769-1775.

Sloan, A.W. et al. (1969).   Nutrition and physical fitness of white, coloured and Bantu high-school children.   Medical Journal, **43**, 508-511.

Spurr, G.B. (1984).   Physical activity, nutritional status and physical work capacity in relation to agricultural productivity,   In: E. Pollitt & P. Amante (eds.), Energy Intake and Activity, pp. 207-261.   New York: Liss.

Stini, W.A. (1975).   Adaptive strategies of human populations under nutritional stress,   In: E.S. Watts, F.E. Johnston & G.W. Lasker (eds.),   Biosocial Interrelations in Population Adaptation.   The Hague.

Sukhatme, P.V. & Margen, S. (1982).   Autoregulatory homeostatic nature of energy balance.   American Journal of Clinical Nutrition, **35**, 355-365.

Twiesselman, F. (1957).   De la croissance des ecoliers noirs de Leopoldville. Academie royale des sciences coloniales, classe des sciences naturelles et medicales, memoire no. 8 Tome VI, 7.

Twiesselman, F. (1965).   Expedition anthropologique du Dr.D.J.H. Nyessen. II. Les Aromo-gallas, les Anouakes, les Chippouks, les Nouers. Bulletin de la Societe Belge d'Anthropologie et de Prehistoire, **75**, 121-162.

Twiesselman, F. (1969).   Developpement biometrique de l'enfant a l'adulte. Maloine, Paris: Presses Universitaires de France.

Vincent, M., Jans, C. & Ghesquiere, J. (1962).   The newborn pigmy and his mother.   American Journal of Physical Anthropology, **20**(3), 237-247.

Viteri, F.E. (1971).   Considerations of the effect of nutrition on the body composition and physical working capacity of young Guatemalan adults, In: N.S. Scrimshaw & A.M. Altschul (eds.),   Amino Acid Fortification of Protein Foods.   Cambridge.

ETHNIC AND SOCIO–CULTURAL DIFFERENCES
IN WORKING CAPACITY

GENETICS OF WORKING CAPACITY

D. F. ROBERTS

*Department of Human Genetics, University of Newcastle upon Tyne,
Newcastle upon Tyne, England*

INTRODUCTION

An outstanding feature of recent human evolution has been change in activity patterns. Today, many of us spend much of our time in sedentary tasks, making little use of the remarkable physical capacities with which our ancestors endowed us. If primitive man did not work, he did not eat, though that work would have incorporated a variety of activities and requirements - agility in climbing a tree for fruits, rapid bursts of speed in hunting and endurance in pursuit of wounded game, size and strength in the final attack on the quarry, and in carrying the proceeds of the hunt or the harvest back to the group. Success in such skills would obviously have been of survival importance, and natural selection would have favoured phenotypes enhancing that success. The genetic component of it, however small, is likely to have been selected for. Yet to establish the extent of that genetic component is extremely difficult.

Factors responsible for this are first the long human generation length, which makes it difficult to obtain comparable records for parents, children and other relatives. Secondly, there is the change in work capacity that comes in the long term with ageing and in the short term with variations in health, nutrition and similarly, and there is the difficulty of distinguishing the effects of intrinsic genetic characters from those due to environment. Then, most important, there is biological complexity itself. It is the almost infinite number of variables that are involved in work capacity that is mainly responsible for the difficulty in any attempt to measure the genetic contribution to it.

It is not surprising, therefore, that there is very little information in the literature relating to the extent of genetic influence on work capacity. There is, however, a little on its components, that is to say on physiological fitness and performance, on adaptation to work and aerobic power, on characteristics of muscle tissue, on muscle strength and on motor performance.   There is rather more on the genetic contribution to body size, physique, and body composition. This perhaps is one way of attacking the problem, dissecting out various contributory components to work capacity, and examining one by one the genetic contribution to them.   Recently, there have been several excellent critical reviews that adopt this approach, e.g. on endurance (Bouchard & Lortie, 1984), on physiological fitness and motor performance (Bouchard & Malina, 1983), and on aerobic power and capacity (Bouchard, 1985).

Genetic studies fall into several categories.   First there are the studies of twins in which the variances between members of dizygotic and monozygotic twin pairs are compared.   One assumption is that monozygotic twins have the same genetic constitution, so that the variance between them is entirely of environmental origin, whereas the variance between dizygotic twins contains the same environmental component as well as a genetic one.   Another is that the environmental influence on monozygotic twins is the same as on dizygotic. Twin studies belie their apparent simplicity, for they have many limitations;   for example, apart from the need to establish zygosity objectively, the initial premise of genetic identity of monozygotic twins is open to question except immediately after the cell division at which twinning occurs soon after fertilisation.   All that twin studies can do is to indicate the possibility that there is some genetic involvement, and cannot give an accurate measure of its extent or of the mode of inheritance.   Family studies are much more useful, for they allow the calculation of correlations between relatives of different degrees and from these the heritabilities, and with modern methods of analysis, employing path coefficients, mixed models of inheritance, and similarly, are particularly informative.   When these family studies can be extended to include adopted individuals, then very valuable distinction of the extent of environmental and biological factors can be obtained.   Needless to say, there are few such studies. "Heritability" calculated from twin studies is not strictly comparable with that calculated from family studies.

Table 1 summarises the results of studies of respiratory function.   Each "+" indicates a study which reports or suggests a positive genetic contribution, each "-" a study showing no significant genetic element.   The majority consist of twin studies which almost consistently show greater variablity between dizygotic than between monozygotic twins.   The effect on vital capacity remains when account is taken of sex, stature, body weight, ethnic origin and the amount of exercise.   The difference in significance of the genetic effect on maximal flow between 60% and 40% of total lung capacity suggests that the genetic

Table 1: Genetic influence in respiratory function

	Twin studies	References	Family studies	References
Vital capacity	+++	Arkinstall et al, 1974; Man & Zamel, 1976; Pirnay et al, 1975		
Vital capacity per unit body weight	+	Engström & Fischbein, 1977		
Total lung capacity	+	Pirnay et al, 1972		
Maximum respiratory volume	+	Cotes et al, 1977		
Forced expiratory volume (FEV)	-	Pirnay et al, 1972		
Forced vital capacity			+	Mueller et al, 1980
FEV 3			+	Mueller et al, 1980
FEV 1			+	Mueller et al, 1980
Peak flow rate			+	Mueller et al, 1980
Maximal flow at 60% total lung capacity	+	Man & Zamel, 1976		
Maximal flow at 40% total lung capacity	-	Man & Zamel, 1976		
Tidal volume response	+	Arkinstall et al, 1974		
Breathing frequency response	-	Arkinstall et al, 1974		
Ventilatory response	-	Arkinstall et al, 1974		
Ventilatory response to hypoxia			+	Scoggin et al, 1978
Ventilatory response to hypercapnia			-	Scoggin et al, 1978
Ventilatory response to progressive isocapnic hypoxia	+	Collins et al, 1978		
Ventilatory response to hyperoxic hypercapnia	-	Collins et al, 1978		

+ = a study which reports or suggests a positive genetic contribution
- = a study which shows no significant genetic element

Table 2:   Genetic influence in work capacity

	Twin studies	Twin studies References	Family studies	Family studies References
Maximal aerobic power (VO$_2$max/kg)	++ -	Kimura, 1956; Komi et al, 1976 Howald, 1976; Klissouras et al, 1973; Komi & Karlsson, 1979	-	Howald, 1976
Maximal working capacity (watts/kg weight)				
PWC$_{205}$	+	Komi et al, 1973		
PWC$_{150}$/kg	-	Bouchard et al, 1982		
Maximum anaerobic muscular power	-	Komi et al, 1973		
Maximum amount of work on bicycle ergometer (6 mins), PWC/leg, max.	+	Engström & Fischbein, 1977		
Maximum amount of work adjusted for amount of habitual exercise	- -	Engström & Fischbein, 1977; Weber et al, 1976		
Maximum aerobic power adjusted for age, body weight, fatness			+	Montoye & Gayle, 1978
Maximum aerobic power adjusted for fatness, socioeconomic status, habitual activity, smoking:				
overall/kg body weight			+	Bouchard, 1982; Lortie et al, 1982
overall/unit SA			+	Bouchard, 1982; Lortie et al, 1982
overall/unit height			+	Bouchard, 1982; Lortie et al, 1982

component in this is related to inherited variation in shape and size of the large airway passages, and this would account for the finding of genetic influence in vulnerablity of the lungs and the airways to various diseases (Collins et al, 1978; Webster et al, 1979). As regards hypoxic and hypercapnic ventilatory responses, the significant genetic variance in tidal volume response but not in breathing frequency response or ventilatory response at three levels of end tidal P_{CO_2} again suggests a relationship to size. Of the few family studies that have been carried out, the correlations between relatives in forced vital capacity, FEV3, FEV1 and peak flow rate, suggest, when the scores are adjusted for age, sex and size, a low heritability of lung functions, with genetic effects highest (30-50%) for peak flow rate and FEV1. The fact that in the study of Mueller et al (1980) the correlations changed on migration to a different altitude indicates the clear environmental influence on these functions. There is acceptable agreement when both types of study have been made, e.g. in the significant genetic effect on ventilatory response to hypoxia, but not to hypercapnia.

Table 2 summarises the results of studies of work and aerobic power. Twin studies here show considerable variation, irrespective of what measure is used. However, some of the inconsistency must be due to sample size and inadequate control of associated factors, for the two twin studies with the largest sample sizes controlling for age and sex (Engstrom & Fischbein, 1977; Bouchard et al, 1982) agree well with the two family studies (Montoye & Gayle, 1978; Lortie et al, 1982) in suggesting that the heritability of maximum aerobic power is not more than 40%. Heritabilities calculated from earlier twin data are shown in Table 3a. Recent work by Bouchard (1985) suggests that whereas the genetic effect in maximum heart rate is quite high (70%), it is much lower (<30%) in ventilatory thresholds, maximal aerobic power

Table 3a: Twin studies of work capacity

	Intraclass correlations		Heritability	Reference
	MZ	DZ	$2(r_{MZ} - r_{DZ})$	
Maximum O_2/kg/min	.91	.44	94	Klissouras et al, 1971
	.95	.36	118	Klissouras et al, 1973
Maximum heart rate	.90	.48	84	Klissouras et al, 1971
	.84	.62	44	Klissouras et al, 1973
Heart volume/weight	.42	.28	28	Klissouras et al, 1973
PWC max/kg, or	.93	.43	100	Komi et al, 1973
PWC_{205} kg	.68	.28	80	Engström & Fischbein, 1977
PWC_{150} kg	.60	.41	38	Bouchard, 1985

Table 3b: Twin correlations and heritabilities
             (Bouchard, 1985)

	MZ	DZ	h^2 %
Maximum aerobic power/kg	.76	.73	6
–ditto–	.80	.74	12
Maximum heart rate	.81	.45	72
VT – 1/kg	.67	.52	30
VT – 2/kg	.64	.52	24
MAC/kg	.79	.63	32
PWC_{150}/kg	.60	.46	28

and capacity, and submaximal power outputs (Table 3b).   Comparisons of populations of similar or different genetic constitution, living under contrasting environmental conditions, are inconclusive with respect to the extent of genetic differences.

Studies of static strength (Table 4) cover a variety of tasks, and there is considerable variation also in the type of measurement used.   These may account for some of the inconsistencies between studies, but on balance the majority of twin and family studies show the existence of low to moderate heritabilities, e.g. of 35-55% for grip, push and pull corrected for body weight and reliability.   Dynamic strength studies show similar variability, with heritability estimates ranging up to almost complete in running velocity (Komi, 1973), but with inconsistencies among studies, e.g. with estimates of 22%, 40% and 85% heritability for push-ups.

Table 5 summarises studies on motor tasks.   Fine motor tasks involving hand steadiness, mirror drawing, tapping speed, dexterity, pursuit tracking and manual coordination, all suggest appreciable genetic effects, with heritabilities ranging from 50 to 80%.   These tend to be higher for performances with the right hand than with the left, and tend to be higher in one sex than the other according to task.   For nerve conduction velocity, reaction time and reflex time, results of twin studies are quite discordant, the highest heritability estimates coming from electroencephalogram traces.   As regards gross motor tasks, agility items seem to have a stronger genetic component than the power items.   Throwing tasks give more variable estimates than jumping tasks, and flexibility tasks a higher genetic component than balance tasks.   An interesting comparison of performance of college males with that of their fathers, a generation earlier, showed correlations ranging from .86 for the running long jump to .04 for the fence vault.   Motor tasks in parents and offspring showed

Table 4: Muscular strength

	Twin studies	References	Family studies	References
Back strength	−	Kimura, 1956	+ −	Wolanski & Kasprzak, 1979 Kovar, 1981
Trunk extension	+++++	Ishidoya, 1957 Kovar, 1981		
Grip strength	+++++	Ishidoya, 1957; Kovar, 1981	++ +	Kovar, 1981 Wolanski & Kasprzak, 1979
Elbow flexion	+++	Kovar, 1981	+	Kovar, 1981
Knee extension	++	Kovar, 1981		
Shoulder strength	+	Ishidoya, 1957	+	Wolanski & Kasprzak, 1979
Composite score (handgrip, knee stretch, arm bend)	+	Engström & Fischbein, 1977		
Composite score (right and left grip, back lift)	+	Mizuno, 1956		
Composite score (arm strength, grip and relative strength index)				
Right grip			+	Montoye et al, 1975
Left grip			+	Montoye et al, 1975
Push			+	Malina & Mueller, 1981
Pull	+	Malina & Mueller, 1981	+	Malina & Mueller, 1981
Isometric quadriceps	−	Komi et al, 1973		
Isometric and isotonic forearm flexors	−	Komi et al, 1973		
Dynamic strength:				
Push up	+ −	Ishidoya, 1957 Kovar, 1981	+	Weiss, 1979
Sit up	+	Kovar, 1975		
Maximum muscular power (running velocity):				
Males	+	Komi et al, 1973		
Females	−	Komi et al, 1973		

Table 5: Motor tasks

Task	Twin studies			Family studies		
		References	Herit-ability, h² %		References	Herit-ability, h² %
Running long jump	–			+	Cratty, 1960	
100 yd run				+	Cratty, 1960	
Chin-ups				–	Cratty, 1960	
Fence vault				–	Cratty, 1960	
Reaction time	+++	Vandenberg, 1962	32			
		Komi et al, 1973; Kovar, 1981; Sklad, 1973	14–86			
Hand movement speed				+	Wolanski & Kasprzak, 1979	
Eye-hand coordination				+	Wolanski & Kasprzak, 1979	
Walking balance				–	Wolanski & Kasprzak, 1979	
Turning balance				+	Wolanski & Kasprzak, 1979	
Turning speed				–	Wolanski & Kasprzak, 1979	
Composite (Johnson test) balance and agility	+		74–80			
Pursuit rotor task	+					
Hand movement accuracy	++	Kimura, 1956; Kovar, 1975, 1981; Marisi, 1977; McNemar, 1933; Sklad, 1973	50–80	–	Wolanski & Kasprzak, 1979	
Tapping speed	+	Sklad, 1973, 1975		+	Kovar, 1981	
Manual coordination time	+			+	Kovar, 1981	
Manual coordination errors	+			–	Kovar, 1981	
Ball rolling accuracy						
Various running tasks	+++	Mirenva, 1935; Ishidoya, 1957; Kimura, 1956; Kovar, 1975	45–48	–	Malina & Mueller, 1981	26
Various jumping tasks	++++	Mirenva, 1935; Ishidoya, 1957; Kimura, 1956; Sklad, 1972	45–48	–	Malina & Mueller, 1981	22
Various throwing tasks	–+	Ishidoya, 1957; Sklad, 1972	14–71	+	Malina & Mueller, 1981	56
Flexibility tasks	++	Kovar, 1975, 1981	48			
Beam balance	+	Vandenberg, 1962	24			
Miles body sway	+	Vandenberg, 1962	46			
Bachman ladder climb		Williams & Hearfield 1973				
Short distance dash	+++	Wolanski et al, 1980	45–91	–	Montoye & Gayle, 1978	

+ = a study which reports or suggests a positive genetic contribution; – = a study which shows no significant genetic element.

low to moderate parent/child correlations. In general, however, many motor tasks show moderate heritabilities, though this result may be somewhat distorted by the absence of consideration of factors of learning, training, habitual activity and practice, maturation, body size and physique.

CONCLUSION

It appears from this review that most of the effort devoted to understanding the genetics of working capacity has been channelled to twin studies. These, and particularly the earlier ones, are mainly based on very small numbers of twins, their zygosity may not have been properly identified by serological methods, the series often include twin pairs covering a range of ages, and some of the interpretations leave much to be desired. The effects of concomitant variables have all too frequently been ignored. The first conclusion is that this field remains wide open for exploration by more sophisticated modern methods of genetic analysis.

That such investigation would be profitable is indicated by the data so far. There is, as expected, very little evidence of any major single gene effect on the variables considered, other than where working capacity is impeded by some simply inherited genetic disease. Instead, any genetic effect that there may be seems to be describable by the multifactorial model, in which the genetic component is polygenic, with the genes exerting additive and non-additive effects within the total phenotypic variance. That is to say, the analysis is to be carried out by standard methods of quantitative genetics. Most twin studies give an indication of a measurable genetic proportion of the variation in respiratory function, working capacity, muscular strength, motor tasks, but varying from one task to another. What is now required is repetition of those investigations in full family studies to assess the extent of genetic control.

Moreover, virtually all studies so far have been made on populations of European derivation, and certainly no attempt has been made to measure the genetic component in any population differences that may exist. Their analysis is much more difficult, requiring investigation of first generation hybrids. It is not impossible, however, and may be particularly rewarding.

REFERENCES

Arkinstall, W.W., Nirmel, K., Klissouras, V. & Milic-Emili, J. (1974). Genetic differences in the ventilatory response to inhaled CO_2. J.Appl.Physiol. **36**, 6-11.

Bouchard, C. (1982). Family resemblance in selected biological traits: a preliminary report. In: R. J. Shephard & H. Lavallee (eds.), Child

Growth and Development/Croissance et developpement de l'Enfant,   pp. 122-130. Trois-Rivieres, PQ: Editions du Bien Publique.

Bouchard, C. (1985).   Genetics of aerobic power and capacity,   In: R. M. Malina (ed.), Human Genetics and Sport (in press).

Bouchard, C. & Lortie, G.   (1984).   Heredity and endurance training.   Sports Medicine, 1, 38-64.

Bouchard, C. & Malina, R.M. (1983).   Genetics of physiological fitness and motor performance. Exercise and Sport Sciences Reviews, 11, 306-339.

Bouchard, C., Leblanc, C., Lortie, G., Simoneau, J.A., Theriault, G. & Tremblay, A. (1982).   Submaximal physical working capacity in adopted and biological siblings. Medical Science and Sports, 14, 139.

Collins, D.D., Scoggin, C.H., Zuillick, C.W. & Weil, J.V. (1978).   Hereditary aspects of decreased hypoxic response. Journal of Clinical Investigation, 62, 105-110.

Cotes, J.E., Heywood, L.C. & Laurence, K.M.   (1977).   Determinants of respiratory function in boy and girl twins.   In: J. S. Weiner (ed), Physiological Variation and its Genetic Basis,   pp. 77-86.   London: Taylor & Francis.

Cratty, B.J.   (1960).   A comparison of fathers and sons in physical ability. Res.Q. 31, 12-15.

Engström, L.M. & Fischbein, S. (1977).   Physical capacity in twins.   Acta Genet.Med.Gemellol., 26, 159-165.

Howald, H. (1976).   Ultrastructure and biochemical function of skeletal muscle in twins. Ann.Hum.Biol. 3, 455-462.

Ishidoya, Y. (1957). Sportfähigkeit der Zwillinge.   Acta Genet.Med.Gemellol. 6, 321-326.

Kimura, K.   (1956).   The study of physical ability of children and youths: on twins in Osaka City. Jinrui Idengaku Zasshi, 64, 172-196.

Klissouras, V. (1971).   Heritability of adaptive variation. J.Appl.Physiol. 31, 338-344.

Klissouras, V., Pernay, F. & Petit, J.M. (1973).   Adaptation to maximal effort: Genetics and age. J.Appl.Physiol. 35, 288-293.

Komi, P.V. & Karlsson, J.   (1979).   Physical performance, skeletal muscle enzyme activities, and fibre types in monozygous and dizygous twins of both sexes. Acta Physiol.Scand. (Suppl.), 462, 1-28.

Komi, P.V., Klissouras, V. & Karvinen, E. (1973).   Genetic variation in neuromuscular performance. Int.Z.angew.Physiol., 31, 289-304.

Komi, P.V., Viitasalo, J.T., Havu, M., Thorstensson, A. & Karlsson, J.   (1976). Physiological and structural performance capacity: effect of heredity, In: P. V. Komi (ed.),   Biomechanics, vol. 5,   pp. 118-123.   Baltire: University Park Press.

Kovar, R.   (1975).   Motor performance in twins. Acta Genet.Med.Gemellol. 24, 174.

Kovar, R.   (1981).   Human Variation in Motor Abilities and its Genetic Analysis. Prague: Charles University.

Kovar, R. (1981). Sledovani podobnosti mezi rodici a jejich potomky v nekterych motorickych projevech. Teorie a Praxe Telesne Vychovy, **29**, 93-98.

Lortie, G., Bouchard, C., Leblanc, C., Tremblay, A., Simoneau, J.A. & Savoie, J.P. (1982). Familial similarity in aerobic power. Human Biology, **54**, 801-812.

McNemar, Q. (1933). Twin resemblances in motor skills, and the effect of practice thereon. J.Genet.Psychol. **42**, 70-99.

Malina, R.M. & Mueller, W.H. (1981). Genetic and environmental influences on the strength and motor performance of Philadelphia school children. Human Biology, **53**, 163-179.

Man, S.F.P. & Zamel, N. (1976). Genetic influence on normal variability of maximum expiratory flow-volume curves. Journal of Applied Physiology, **41**, 847-877.

Marisi, D.G. (1977). Genetic and extragenetic variance in motor performance. Acta Genet.Med.Gemellol. **26**, 197-204.

Mirenva, A.N. (1935). Psychomotor education and the general development of preschool children: experiments with twin controls. J.Genet.Psychol. **46**, 433-454.

Mizuno, T. (1956). Similarity of physique, muscular strength and motor ability in identical twins. Bulletin of the Faculty of Education, Tokyo University, **1**, 190-191.

Montoye, H.J. & Gayle, R. (1978). Familial relationships in maximal oxygen uptake. Human Biology, **50**, 241-249.

Montoye, H.J., Metzner, H.L. & Keller, J.K. (1975). Familial aggregation of strength and heart rate response to exercise. Human Biology, **47**, 17-36.

Mueller, W.H., Chakraborty, R., Barton, S.A., Rothammer, F. & Schull, W.J. (1980). Genes and epidemiology in anthropological adaptation studies: familial correlations in lung function in populations residing at different altitudes in Chile. Med.Anthropol. **4**, 367-384.

Pirnay, F., Klissouras, V. & Petit, J.M. (1972). Function pulmonaire et genetique. Acta Tuberc.Pneumol.Belg. **6**, 477-483.

Pirnay, F., Klissouras, V., Deroanne, R. & Petit, J.M. (1975). Aptitude physique des jumeaux. Medicine & Sport, **49**, 29-33.

Scoggin, C.H., Dockel, R.D., Kryer, M.H., Zwillick, C.W. & Weil, J.V. (1978). Familial aspects of decreased hypoxic drive in endurance athletes. Journal of Applied Physiology, **44**, 464-468.

Sklad, M. (1972). Similarity of movements in twins. Wychowanie Fizyczne i Sport, **16**, 119-141.

Sklad, M. (1973). Rozwoj fizyczny i motorycznosc blizniat. Materially i Prace Antropologiczne, **85**, 3-102.

Sklad, M. (1975). The genetic determination of the rate of learning of motor skills. Studies in Physical Anthropology, **1**, 3-19.

Vandenberg, S.G. (1962). The hereditary abilities study: hereditary components in psychological test battery. Amer.J.Hum.Genet. **14**, 220-237.

Weber, G., Kartodihardjo, W. & Klissouras, V. (1976). Growth and physical training with reference to heredity. Journal of Applied Physiology, **40**, 211-215.

Webster, P.M., Lorimer, E.G., Man, S.F.P., Woolf, C.R. & Zamel, N. (1979). Pulmonary function in identical twins. American Reviews of Respiratory Disease,. **119**, 223-228.

Weiss, V. (1979). Die Heritabilitäten sportlicher Tests, berechnet aus den Leistungen zehnjäten Zwillingspaare. Leistungssport, **9**, 58-61.

Williams, L.R.T. & Hearfield, V. (1973). Heritability of a gross motor balance task. Research Quarterly, **44**, 109-112.

Wolanski, N. & Kasprzak, E. (1979). Similarity in some physiological, biochemical and psychomotor traits between parents and 2-45 years old offspring. Studies in Human Ecology, **3**, 85-131.

Wolanski, N., Tomonari, K. & Siniarska, A. (1980). Genetics and the motor development of man. Human Ecology & Race Hygiene, **46**, 169-191.

WORKING CAPACITY IN DIFFERENT AFRICAN GROUPS

J. HUIZINGA

Institute of Human Biology, State University at Utrecht,
The Netherlands

INTRODUCTION

The Utrecht group started its programme on working capacity and related parameters in African populations towards the end of the I.B.P. period. Part of this initial work has been published (Huizinga & Reijnders, 1974).

The population concerned was a group of 86 males, mainly consisting of members of the agricultural Fali, a Kirdi-tribe living in the mountainous region of North Cameroun between the 8th and 10th degree North latitude. They were studied in the neighbouring villages of N'Goutchoumi and Hama Koussou during the early months of 1970, i.e. during the dry season. The age of these volunteers ranged between 16 and 47 years.

As an ergometer test, the double Master's step test has been chosen, as described in the I.B.P. handbook (Weiner & Lourie, 1969, p. 226). Heart frequency was determined at the ictus by auscultation; the number of beats was counted during 20 seconds immediately after each three-minute stepping period. No attempt was made to determine oxygen consumption directly. Instead, the 1954 nomogram of Åstrand and Ryhming was used to predict maximum aerobic power. General medical examination resulted in disqualification of eight individuals. Another 26 people performed badly and their results were discarded. Thus, the resulting 52 satisfactory tests, discussed previously (Huizinga & Reijnders, 1974) may be considered to represent those of selected males. Ever since, the relatively high values of aerobic power found (about 3.5 l/min and some 60 ml/kg/min) puzzled us. In Shephard's survey (1978, Table 17) the highest means in Africans published at that time concerned active Tanzanians (57.2 ml) and active Yoruba (55.5 ml). Moreover, our later personal experiences in several more African groups seemingly living at more or less the same daily activity level as the Fali showed much lower values. Recent reconsideration of the original Fali data revealed that the rules in applying the nomogram had not been rigidly adhered to, in several cases the heart rate associated with the highest load, i.e. the highest stepping rate, had remained

Table 1:   Maximum aerobic power based on auscultatory determined post-exercise heart frequencies

Population	Sex	Year Studied	Age (yrs)	n	VO$_2$max* l/min	ml/kg/min
Fali						
Farmers	Male	1970	18-34	30	3.11 ± 0.49	55.4 ± 9.1
			>35	8	3.14 ± 0.56	56.1 ± 10.2
Sudanese						
Soldiers	Male	1974	21-31	10	2.92 ± 0.51	46.4 ± 10.2
Students	Male	1974	20-29	14	2.57 ± 0.39	42.2 ± 8.5
Nurses	Female	1974	20-24	20	2.03 ± 0.35	38.9 ± 8.1

*   Åstrand nomogram (1960)

below 120 bpm.   As a low heart rate at a high load leads to a high predicted maximum aerobic power the solution to the surprising Fali data seemed near. The, as yet, unpublished recalculated means (now based on the Åstrand nomogram) of the remaining 38 males, indeed appeared to be lower than those originally published.   However, the mean figures (Table 1) are still high enough to be remembered: 3.11 ± 0.49 l/min or 55.4 ± 9.1 ml/kg/min in the younger people (18-34 years; n=30), and 3.14   0.56 l/min or 56.1 ± 10.2 ml/kg/min in the >35 years group (n=8).

Auscultatory post-exercise heart frequency determination has also been applied during our work in Khartoum in 1974.   In addition to the Fali data these as yet unpublished data on maximum aerobic power are given in Table 1.   In terms of African ethnic differences this table does not reveal spectacular differences: the maximum aerobic power in well-nourished and physically trained Sudanese soldiers is on the average of the same magnitude as that found in the Fali (about 3 l/min); however, in the heavier soldiers, (≃ 63 kg as against ≃58 kg in the Fali) the relative VO$_2$max remains on the average well below (46.4 ml) the Fali value (55.4 ml).   During the dry season, the Fali people nutritionally also were in good condition.

The usual sex difference in maximum aerobic power is also present in the data given in Table 1.   It will appear later that the well-nourished Sudanese nurses from Khartoum do not perform markedly better than their sisters from small West-African villages.

A problem for step test users is met the moment that one wants to calculate the loads at the different stepping rates.   Of course, the almost equidistant stepping rates (54, 76, 100, 120) may be considered to represent relative loads in an individual case of constant weight.   Evidently, the method used to calculate load is decisive in the evaluation of ethnic or other differences as determined by different investigators.   In their original publication on the 'two-step test', Master and Oppenheimer (1929) reasoned that "on the descent, there is no actual lifting of the weight against gravity. Since any work done in descending will be comparatively small in amount, and exactly proportional to the work done on the ascent, for the sake of simplicity this is omitted in the calculation".   The calculation concerned the work performed per minute, i.e. in the original version the product of body weight and the total vertical distance the body is raised during exercise.

History has taught that 'for the sake of simplicity' is not a good enough reason to apply this formula in inter-individual or inter-populational research. Despite the fact that Shephard and Olbrecht (1970) stressed that there is indeed little need to adjust for differences in body weight because at a consistent stepping rate almost all individuals are subjected to the same relative work load some doubts remained.   Also Lange Andersen et al (1971) advocate that to calculate the work performed on a step test simply as the product of body weight (kg), height of step (m) and number of ascents in a certain time, but then they have to conclude that "a load of 700 kpm/min on the step test is approximately equivalent to a load of 1000 kpm/min on the bicycle ergometer".   This difference is not explained away easily, unless one accepts that *descending* accounts for about 30% of the total work performed on a step test.   This also follows from the nomogram of Åstrand and Ryhming (1954) when one compares the work loads of the step test and bicycle ergometer respectively.   Also others (e.g. Pudelski, 1968) found experimentally that descending accounts for an appreciable percentage (22-28%) of the total work performed (W_T).   In the light of the widely different estimates of the relative contribution of descending to the total work load we decided, irrespective of stepping rate, to use on purely arithmetic grounds (see formula), an estimate of 31%.   Hence, the work during ascents (W_{asc}) amounts to 0.69 W_T.   Thus, the total work W_T amounts to $\frac{100 \times W_{asc}}{69}$. As W_{asc} (kg/m/min) equals the continuous product of the number of ascents/minute (N_{asc}), stepping height (0.46 m) and body weight (Wt; kg) W_T may be calculated as

$$W_T : \frac{100}{69} \times N_{asc} \times 0.46 \times Wt, \quad \text{or} \quad W_T = \frac{2}{3} \times N_{asc} \times Wt.$$

In the two-step test applied, the stepping rhythm is set by the metronome at 54, 76, 100 and 120 bpm.   Six metronome beats accompany *one ascent* plus one descent per minute, i.e. 54 bpm stands for 9 ascents, etc.   Substitution in the

Table 2:   Relative load and mean heart rate in the Bozo (Mali, 1977)

fc (b/m)	Cardiotachometry n = 64	Auscultation n = 46
f_0	75 ±12.3 (16.4)*	76 ±12.7 (16.7)
f_{54}**	105 ± 10.8 (10.3)	89 ±16.3 (18.3)
f_{76}	117 ± 11.8 (10.1)	107 ± 16.0 (15.0)
f_{100}	134 ± 13.8 (10.3)	126 ± 15.1 (12.0)
f_{120}	154 ± 14.6 (9.5)	145 ± 16.1 (11.1)

*   Coefficient of variation (%)
**   Pace = 54 metronome beats/min (etc.)

above formula results in a simple expression of the relationship between load, stepping rate and body weight.   At the respective rates, the *loads* in kpm/min amount to 6 x Wt, 8.5 x Wt, 11.1 x Wt, and 13.3 x Wt, respectively.   These formulae have been used in all our subsequent work with the step test.   However, we did change the stethoscope to count the heart beats for a continuously recording cardiotachometer (CTM).   As compared to the stethoscope and stop watch method the use of the CTM results in most cases in substantially *higher* post-exercise heart frequencies.

In our Bozo study (Huizinga & Hooijen-Bosma, 1978; Huizinga, 1979) both methods of determination of the heart frequency immediately after each period of three minutes stepping were applied to the majority of individuals.   In Table 2 the resulting differences in mean frequencies at several relative loads (represented by the respective stepping rates) are given. It is clear that, except during the resting state, not only the means but also the respective coefficients of variation differ systematically.   Analysis of the two regression lines (CTM: y = 0.75x + 62; Ausc: y = 0.81x + 46) reveals that only the difference between the intercepts (16 bpm) is significant.

Generally, the Bozo, like the Fali also live in the northern semi-arid zone (thorn savanna belt) between the 12° and 15°N latitude.   However, the Bozo live in villages and some larger settlements (like the small town of Djenne) located in the so-called interior delta of the river Niger and especially in the Bani-Niger region in the republic of Mali.   They are considered to represent one of the oldest populations living in this area.   These 'masters of the river' are notorious fishermen, who are always to be found along or on the river with their 'pirogues'. The availability of water (vegetable gardens; rice, millet, fish) throughout the year contrasts with the situation of other agricultural peoples as the irregular

Table 3: Maximum aerobic power in the Bozo (Mali, 1977)
(Double Master step test; CTM-data)

Age group (yrs)	n	VO_2max* l/min	ml/kg/min
18-24	20	2.59 ±0.43	43.5 ± 6.5
25-34	19	2.89 ±0.79	44.6 ±12.3
>35	16	2.88 ±0.63	42.8 ± 5.0

* Åstrand nomogram (1960)

distribution of rainfall during the year makes the food supply in the latter groups more vulnerable. The Bozo are the only case in which Pales et al (1953) in their 'catalogue' of West-African populations refer to an "excellent nutritional state" and a "great vigour" (p. 257), apparently in their experience not the most obvious characteristics of human groups in this part of the world.

Our data on working capacity of Bozo males as determined in 1977 by means of the step test with CTM-heart frequency determination do not substantiate the view that these people would occupy a very exceptional position (Table 3).

Most of our work on predicted maximum aerobic power (Åstrand, 1960) in African males after 1977 has been done with a bicycle ergometer. Three examples may be given.

In December and January 1978/79 we studied in Djenne, the small town where also the Bozo were found, a number (n=52) of 'non-Bozo' adult Djenneans (Huizinga, 1982). The Swedish mechanically-braked Monark bicycle was used in this instance. It may be remarked also that the Åstrand (1960) nomogram has been constructed on the basis of Monark data. In the Monark tables, allowance is made for the so-called 'frictional losses' which, in accordance with the estimate of Cumming & Alexander (1968) amount to about 9%. We hoped by the use of these tables to further the possibility of direct comparison of data obtained by other means (step test and the electrically-braked Lode bicycle as used in subsequent studies).

The mode of living of the average Djennean is heavily influenced by the presence of the river; their diet is similar to that of the Bozo. Physically they do not differ appreciably from Bozo males of comparable age: on the average they tend to be slightly heavier (64.8 kg as against 63.3 kg), the difference being allowed for by musculature (LBM 56.1 kg against 54.8 kg), and somewhat taller stature (172.3 cm versus 171.0 cm).

198   *Huizinga*

Table 4:   Maximum aerobic power in Djenneans (Mali, 1978/79)
           (Bicycle ergometer; Monark)

Age group (yrs)	n	VO$_2$max* l/min	ml/kg/min
18-24	20	2.67 ±0.68	42.5 ± 8.9
25-34	21	2.62 ±0.64	39.8 ± 7.2
>35	9	2.51 ±0.85	37.0 ±10.6

*   Åstrand nomogram (1960)

In Table 4 the data under discussion are given. In each of the three age groups the 'average Djennean' as compared to the Bozo tends to show a lower relative maximum aerobic power, especially in the males above 25 years of age. This systematic difference may well derive from differences in daily activity within the community of Djenne.   Non-Bozo also are frequently found on the river, though agriculture is their main activity, whereas fishing, sometimes at long distances, forms the main activity of the Bozo.   We do not possess quantitative data on energy expenditure of specified occupations in Djennean people.

In two small villages in Burkina Faso (Upper-Volta) our comparative study of working capacity was continued in 1980 and 1981.   Adult males from both places were subjected to submaximal loads on the Lode electrically-braked bicycle ergometer.

The village of Ziou, some 40 km south-east of the township of Po in the south, near the Ghanean border, has been chosen for an epidemiological study into some depth within the framework of the W.H.O.-Onchocerciasis Control Programme (Huizinga, 1980).   This village was found to be hyperendemic for onchocerciasis: in people of 15 years and older the prevalence amounts to about 100%.   It may be remarked that in the area concerned, four other major parasitic diseases also occur in appreciable frequencies (filariasis Bancrofti, malaria, schistosomiasis, ankylostomiasis).   The present study (January-February 1980) forms a modest beginning of a complex study of the possible relationship between onchocerciasis and working capacity.   The region concerned is now practically under control as far as the Simulium fly is concerned and the present youngsters may be considered not to be affected.   The people identify themselves as Nankana, a small group of farmers living amongst Gourounsi-speaking groups.

Out of 97 adult males 20 were medically disqualified (20%) and did not perform the ergometer test.   As compared to the Fali, where about 10% were

Table 5: Maximum aerobic power in Nankana and Mossi males (Burkina Faso)
(Bicycle ergometer; Lode)

Population	Year Studied	Age (yrs)	n	VO₂max* l/min	VO₂max* ml/kg/min
Nankana					
Farmers	1980	18-24	13	2.30 ± 0.46	40.5 ± 7.8
		25-34	15	2.32 ± 0.48	36.6 ± 6.0
		>35	33	2.10 ± 0.29	34.8 ± 4.6
Mossi					
Farmers	1981	18-24	11	3.00 ± 0.45	52.6 ± 7.1
		25-34	18	2.86 ± 0.55	46.6 ± 6.2
		>35	55	2.11 ± 0.47	36.8 ± 6.4

* Åstrand nomogram (1960)

disqualified on the basis of the same criteria, the Nankana figure seems rather high.   Another 16 males performed badly, i.e. ergometric data are only available on the remaining 61 adult males.   In a second village, Kougpaka (census 1975: 533 inhabitants), 102 males >18 years were examined (January-February 1981). This village is located in the semi-circular hypoendemic onchocerciasis belt some 50 km south of Ouagadougou (Huizinga, 1981).   In people of 15 years and older the prevalence amounts to 32%.   From the ethnic point of view these people belong to the Mossi, a widely dispersed group comprising 48% of the population of Burkina Faso (census 1960-1961).   Thirteen percent of the volunteering males were medically disqualified, i.e. disqualified as far as participation in exercise tests is concerned.   Data on working capacity are available on 84 Mossi males. They are presented in Table 5, together with similar data on the Nankana males. Especially the younger Mossi males (<35 years) perform significantly better than the Nankana.   Both villages are almost 100% dependent on agriculture, most of the products being used to meet their own requirements.   Animal husbandry plays a minor role.   At the time of our studies the Mossi village of Kougpaka especially impressed as a really primitive community without medical or educational facilities.   In these respects, Ziou were slightly better off, with a primary school and a dispensary.   In terms of distances to larger centres, Ziou seem to be more isolated.   Except for onchocerciasis the pattern of disease most probably does not differ much between the two locations.   Of course, we hesitate to attribute the relatively low working capacity as found in the Nankana males to one single factor, i.e. the 100% onchocerciasis infestation.   It should be noted, however, that the Nankana males from Ziou show by far the lowest values

Table 6:   Maximum aerobic power in Nankana and Mossi females (Burkina
                Faso). (Double Master step test; CTM-data)

Population	Year Studied	Age (yrs)	n	VO_2max* l/min	ml/kg/min
Nankana					
Farmers	1980	18-24	16	2.00 ± 0.29	37.5 ± 5.4
		25-34	31	2.06 ± 0.24	37.1 ± 5.4
		>35	17	2.13 ± 0.36	41.2 ± 6.9
Mossi					
Farmers	1981	18-24	20	2.08 ± 0.40	41.3 ± 5.3
		25-34	30	2.04 ± 0.28	40.1 ± 4.7
		>35	42	1.59 ± 0.25	33.4 ± 4.1

*   Åstrand nomogram (1960)

of relative maximum aerobic power (ml/kg/min) in our total male sample
comprising, in addition to the Mossi, also the Fali, Bozo, Djenneans and
Sudanese. In our discussion of the possible significance of anaemia we will again
refer to the Nankana-Mossi difference in working capacity.

The Ziou-Kougpaka difference in relative maximum aerobic power in
males is also found in the case of the younger females. The women chose to
perform the double Master step test instead of the bicycle ergometer. From the
Kougpaka (Mossi-) women, 19% of those volunteering were medically
disqualified. From the initial number of 140 women >18 years of age, 92
performed the step test satisfactorily. In the case of the Nankana women (Ziou)
30% of the volunteering females were excluded from the exercise test; out of
105 initially available women the test was completed by 64 women. It may be
concluded that both in males and females the percentage of people that appeared
to perform the exercise test satisfactorily was much lower in the village of Ziou.

It is clear from Table 6 that, except in the case of the Mossi women >35
years who score unexpectedly low, the two younger female Mossi groups show, on
average, better oxygen transporting systems compared to the Nankana women.
It may be concluded that both Nankana male and female adults show the lowest
mean value of relative maximum aerobic capacity (ml/kg/min) in our total
material.

A perhaps interesting observation was made during the analysis of the
step test data in the Nankana females. Some characteristics of those who
completed the step test were compared to those women in whom the highest

Table 7: Some physical characteristics of Nankana females completing and not completing the step test

	Ht (cm)	Wt (kg)	LBM (kg)	LLV (L)	F (%)	F (kg)
All women (64)	159.7	54.0	39.4	4.0	26.7	14.4
Test completed (46)	159.8	53.5	39.3	4.0	26.2	14.0
Test not completed (18)	159.5	55.3	39.6	4.0	28.1	15.6

stepping rhythm was too strenuous an effort. The drop-outs show on average the same age, the same resting heart frequencies, and the same Hb-concentration as those who completed the test. Physically, only one difference was noted (Table 7): the mean difference of 1.8 kg in body weight could be almost fully accounted for by the difference in fat content (1.6 kg), the drop-outs being the somewhat fatter women. The significance of this observation is still uncertain.

Irrespective of its etiology, low haemoglobin concentrations are generally considered to influence working capacity negatively. Opinions differ, however, as to the kind of relationship between these two variables. Some people favour a 'linear-relationship-model', we ourselves tend to support a 'threshold-model'. Our observations in favour of the latter model were given in Huizinga and Hooijen-Bosma (1980). In the development of the model it was postulated that in anaemics during exercise not only is the increased heart rate the main compensatory factor to allow for the increased demand for oxygen-carrying haemoglobin, but that also other factors leading to a concomitantly increased oxygen-extraction factor play an equally important role. In our model, both males and females appear to show the same threshold: at Hb-values lower than 13.5 g/100 ml the relationship between Hb-concentration and working capacity becomes linear.

If this point of view is accepted it is clear that the number of individuals with an Hb-concentration <13.5 provides an indication of the mean population working capacity to be expected. Obviously, this number is not only determined by the population Hb-concentration mean but also by its standard deviation. One example may be given: the difference in relative aerobic power between the Mossi and the Nankana males indicated a better performance of the Mossi (Table 5). Their respective mean Hb-concentrations do not differ to such an extent as to 'explain' the difference in working capacity: 13.5 g/100 ml in the case of the Nankana (n=97) and 14.1 g/100 ml in the case of the Mossi (n=102).

However, the standard deviation in the Nankana appeared to be twice that found in the Mossi: 2.04 as against 0.99 g/100 ml. The percentage of individuals with Hb-concentrations below the critical value of 13.5 g/100 ml amounts to 38% in the Nankana and to 20% in the Mossi. This observation may well explain the difference in the respective population means as regards maximum aerobic power. The Mossi and Nankana females show much the same pattern as found in the males of these groups: the lower working capacity of the Nankana females is associated wth a somewhat lower mean Hb-concentration and a larger standard deviation (12.1 ± 1.68 g/10 ml versus 13.0 ± 1.11 g/100 ml). These figures result in frequencies of women with Hb 13.5 g/100 ml of 78% (Nankana) and 66% (Mossi) respectively. It is evident that at mean Hb-values of <13.5 a larger standard deviation tends to favour the number of people in which the Hb-concentration surpasses the critical value towards higher concentrations.

In this necessarily brief survey of some aspects of our biomedical work in Africa only observations on adults during dry seasons have been used. Seasonal differences and the observations on schoolchildren remained unmentioned. They will be presented elsewhere.

ACKNOWLEDGMENTS

The cooperation with the World Health Organisation's Onchocerciasis Control Programme (Ouagadougou) during our 1980-1984 work in Burkina Faso greatly stimulated the line of research outlined above.

In particular the elaborate 1980 studies on the Nankana were possible due to a generous grant from Professor H.J.M. Weve Stichting (The Hague), a Dutch foundation involved in combatting blindness in developing countries. The Boise Fund (Oxford) supported our work whenever emergencies arose throughout the programme. The 1982 study was supported by the Wenner-Gren Foundation for Anthropological Research, Inc. (New York).

REFERENCES

Åstrand, I. (1960). Aerobic work capacity in men and women with special reference to age. Acta Physiologica Scandinavica, **49** (Suppl. 169).

Åstrand, P.O. & Ryhming, I. (1954). A nomogram for calculation of aerobic capacity (physical fitness) from pulse rate during submaximal work. Journal of Applied Physiology, **7**, 218-222.

Cumming, G.R. & Alexander, W.D. (1968). The calibration of bicycle ergometers. Canadian Journal of Physiology, **46**, 917-919.

Huizinga, J. (1979). Human biology of the Bozo, fishermen of the river Niger (Mali). Antropologia Contemporanea (Trieste), **2**, 165-189.

Huizinga, J. (1980).   Report on a pilot study (Ziou; Upper Volta), January-February 1980. Report to WHO, Ouagadougou, pp. 9. Utrecht.

Huizinga, J. (1981).   Tropical disease and productivity: the case of onchocerciasis.   January-February 1981, Kougpaka. Progress report to WHO, Ouagadougou, pp. 9. Utrecht.

Huizinga, J. (1982).   Habitual energy expenditure in human populations.   Some methodological implications,   In: N. Wolanski and A. Siniarska (eds.), Ecology of Human Populations, pp. 345-396. Wroclaw: Ossolineum.

Huizinga, J. & Hooijen-Bosma, E.G. (1978).   Predicted aerobic power in the Bozo; experiences with the step test in Mali (West Africa).   Collegium Antropologicum (Zagreb), **2**, 179-188.

Huizinga, J. & Hooijen-Bosma, E.G. (1980).   Hemoglobin concentration and physical fitness,   In: M. Ostyn, G. Beunen & J. Simons (eds.), Kinanthropometry II.   International Series on Sport Sciences, vol. 9, pp. 474-483.

Huizinga, J. & Reijnders, B. (1974).   Heart rate changes during exercise (step test) among the Fali of North Cameroun. Proc.Kon.Ned.Akad.Wetensch, C77, **3**, 283-295.

Lange Andersen, K., Shephard, R.J., Denolin, H., Varnauskas, E. & Masironi, R. (1971).   Fundamentals of exercise testing.   World Health Organisation, Geneva, pp. 133.

Master, A.M. & Oppenheimer, E.T. (1929).   A simple exercise tolerance test for circulatory deficiency, with standard tables for normal individuals. American Journal of Medical Science, **233**, 233.

Pales, L. & Tassin de Saint-Pereuse, M. (1953).   Raciologie comparative des populations de l'Afrique occidentale.   V. Stature;   indice cormique; indice cephalique.   Bull.Mem.Soc.d'Anthrop., Paris, **4**, Serie 10, 3-4, 183-497.

Pudelski, J. (1968).   Evaluation of the magnitude of work in the step test. Annals of the Medical Section of the Polish Academy of Sciences, **13**, 1, 109-133.

Shephard, R.J. (1978).   Human physiological work capacity.   I.B.P. Handbook no. 15. Cambridge University Press.

Shephard, R.J. & Olbrecht, A.J. (1970).   Body weight and the estimation of working capacity. South African Medical Journal, **44**, 296-298.

Weiner, J.S. & Lourie, J.A. (1969).   Human Biology: A Guide to Field Methods. I.B.P. Handbook no. 9. Oxford: Blackwell.

ENVIRONMENTAL, GENETIC AND LEG MASS INFLUENCES ON ENERGY EXPENDITURE

S. SAMUELOFF

*Department of Physiology, H.U. Hadassah Medical School,
Jerusalem, Israel.*

The effects of physical environment, genetic background and muscle mass volume of the extremities on energy expenditure of humans at rest and at maximum work capacity is an intricate physiological problem. Human energy turnover comprises the continuous anabolic and catabolic processes in which transformation of chemical, mechanical, electrical and thermal energies takes place, part of them being transformed to heat. Thus metabolic heat is produced which is the basic mechanism for maintaining body temperature at its normal level. Exposure to heat and cold provokes thermoregulatory reactions like vasodilatation or vasoconstriction, shivering or sweating, while body metabolism is affected as well. Much useful information regarding thermoregulation has been accumulated in populations living in extreme cold like Eskimos, or in primitive conditions with apparently inadequate protection, such as the Australian Aborigines. It is interesting to recall that while the Australian Aborigine can sleep naked through a moderately cold night without recourse to raising his body metabolism and therefore his body temperature can fall to about 35°C, the arctic Eskimo maintains normal body temperature when exposed to cold at the expense of greater metabolic heat production, being unable to tolerate a fall of his body temperature.

In hot climates when the impact of environmental heat overrides the heat dissipation capacity of the human body, body temperature tends to rise. Under these circumstances lowering of metabolic heat production would alleviate the temperature rise. This could be particularly advantageous during exercise when thermoregulation is aggravated by increased metabolism and heat build up. Although this problem has been investigated in laboratory conditions, the conclusions regarding the influence of environmental heat on genetically dissimilar populations at rest and work are still controversial. The controversies seem to stem to a great extent from the multitude of factors which may affect energy exchange in man.   For example, factors such as habitual activity, nutritional and health status, socioeconomic conditions of life or genetic constituents of the examined populations are very often difficult to be critically controlled or even evaluated.   However, these difficulties can be overcome in situations where genetically contrasting populations live in similar climatic and social environments, are engaged in similar occupations, and are exposed to reasonably similar sociopolitical standards of life. In this respect, Israel provides attractive research opportunities to human biologists.

Israel is a small country, but comprises four distinct climatic zones: (1) the strip of coastal area with its mild Mediterranean type of climate; (2) the semi-desert and desert areas of the south, the Negev, which extends to the Sinai peninsula; (3) the highland areas around Jerusalem and Galilee; and (4) the tropical region of the Jordan valley.

The large-scale immigration since the State was established in 1948 brought to Israel Jewish people from many different communities and habitats of the world.   Now one can find in the different climatic zones of the country villages inhabited by people of similar and different ethnic origins.   Extensive genetic studies carried out in the sixties and seventies have found that in many cases there are considerable genetic differences between the various ethnic groups.   One could assume that the genetic differences between these populations have been formed in the adaptation process to the environment in their country of origin.   Demographic surveys have provided information about similarities and dissimilarities of their life patterns.   Altogether, the socio-geographical conditions under which various ethnic groups are settled today in the country offers an unusual opportunity to investigate the relative significance of genetically and environmentally determined variables on physiological characteristics which facilitate life in hot environments.

The objectives of this presentation are to bring together results which we have accumulated during the last few years regarding testing energy exchange and maximum work capacity in summer and winter in subjects of different ethnic origin, inhabitants of two distinctly different climatic zones of the country - the hot south (Negev) and the highland region around Jerusalem.   All subjects were examined in standard laboratory controlled climatic conditions after having achieved a natural acclimatisation to the summer heat in the south or on the Jerusalem hills, and to the winter cold in both areas.   It seems to us that the examiners and experimental set-up of this study provided the conditions for an objective assessment of the significance of inherited and acquired physiological features on adaptation of populations to hot environments.

All subjects in these studies were healthy, young men and women, aged 20-30 years. They were tested in a temperature-controlled room at 24°C and 40% relative humidity. They were examined in summer and this was repeated during the winter. The tests were carried out at least six hours postprandial, and after 1/2 hr supine rest. The population groups studied were Jews of Yemenite origin, who came from the desert-areas of Yemen, Jews of Kurdish origin, who came from the mountainous areas of the Iranian part of Kurdistan, European Jews of Ashkenazi origin and local Bedouins. Throughout this presentation the abbreviated terms - Bedouins, Yemenites, Kurds and Ashkenazim - are used. While the Ashkenazim and the Bedouins are a rather heterogeneous population, there is demographic evidence regarding the Kurds and Yemenites that in their loci of origin, before immigrating to Israel in 1950-52, they had lived for a very long time in virtual isolation from other Jewish communities and therefore they can be regarded as genetic isolates. This supposition has been supported by genetic studies in which differences in blood hereditary factors were disclosed - the Yemenites demonstrated a close resemblance to the Yemenite Arabs, while the Kurdish Jews, although showing a fair resemblance to the indigenous Kurds of Iran, differed markedly in having higher incidence of glucose-6-phosphate dehydrogenase deficiency and β-thalassemia trait in their red blood cells.

A. Resting metabolism was examined in 135 persons (95 men and 40 women). In the Negev, 24 (18 men and 6 women) Yemenites, 10 Bedouin men, 20 Kurds (15 men and 5 women) and 35 Ashkenazi (22 men and 13 women) were tested. In the Jerusalem area the examinees consisted of 20 Yemenites (9 men and 11 women), 16 Kurds (11 men and 5 women) and 10 Ashkenazi men. Resting metabolism was determined by measuring respiratory gas exchange with an open-circuit system and gas collection for five minutes at normal breathing; ECG recordings provided heart-rate data.

No significant differences in resting metabolism were found between the subjects of the various ethnic groups. However, during the summer, the resting metabolism of all subjects (except the Kurdish women) was lower than during the winter, both when resting metabolism was expressed either per body surface area or per Kg body weight (Tables 1 & 2). A slight gain in weight was observed in the winter data of the re-examined subjects. As metabolism was expressed per Kg body weight the observed seasonal changes in resting metabolism could not be due to the winter weight gain.

The observed seasonal variations in resting metabolism were found in the

Table 1:   Seasonal variations in resting metabolism of subjects of different ethnic origin

| | O_2 consumption (ml/m^2 BSA/min ± S.E.) | | | |
| | Males | | Females | |
	Summer	Winter	Summer	Winter
Negev area				
C. Europe	148.0 ±4.4	166.0 ±5.2	129.6 ±4.5	145.5 ±5.0
Yemen	174.6 ±4.4	184.0 ±5.4	156.1 ±5.8	164.8 ±6.5
Kurdistan	166.5 ±5.6	176.6 ±4.1	152.8 ±5.1	154.7 ±5.1
Bedouin		172.1 ±8.1		
Jerusalem area				
C. Europe		164.8 ±6.2		
Yemen	134.7 ±8.9	160.7 ±5.6	148.6 ±5.3	149.6 ±3.4
Kurdistan	153.2 ±6.6	160.2 ±5.8	145.8 ±8.0	141.0 ±8.0

Table 2:   Seasonal variations in resting metabolism of subjects of different ethnic origin

| | O_2 consumption (ml/Kg BW/min ± S.E.) | | | |
| | Males | | Females | |
	Summer	Winter	Summer	Winter
Negev area				
C. Europe	3.9 ±0.1	4.3 ±0.2	3.6 ±0.2	4.1 ±0.1
Yemen	4.6 ±0.1	4.9 ±0.2	4.6 ±0.3	4.8 ±0.2
Kurdistan	4.5 ±0.1	4.8 ±0.1	4.2 ±0.3	4.2 ±0.1
Bedouin		4.8 0.2		
Jerusalem area				
C. Europe		4.2 ±0.2		
Yemen	4.0 ±0.3	4.7 ±0.2	4.3 ±0.2	4.4 ±0.2
Kurdistan	4.0 ±0.2	4.2 ±0.2	3.9 ±0.3	3.7 ±0.3

examinees of all four ethnic groups and it could be assumed that these changes are related to the seasonal variations in environmental heat and the state of natural acclimatisation of the subjects to it.   It should be emphasised that throughout this study all subjects were examined at standard, thermoneutral conditions (24°C).   Therefore, the observed fall in resting metabolism during the summer indicates that natural acclimatisation to summer heat lowers endogenously produced metabolic heat.   It could be assumed that the functional significance of the observed changes is an increased capacity of the body to accumulate heat during the summer when exposed to heat loads.

B.   Work capacity measurements were carried out on 169 subjects, comprising 60 Yemenites, 64 Kurds, 10 Bedouins and 35 Ashkenazi.   All Yemenites and Kurds were tested during the summer and 42 of them were repeated during the winter also.   The Bedouins and Ahskenazi were examined during the winter only. Regarding the location of subjects' settlements, the partition was as follows: from the Negev came 30 Yemenites (14 men and 16 women), 31 Kurds (19 men and 12 women), and 10 Bedouins.   From the Jerusalem highlands came 30 Yemenites (14 men and 16 women), 33 Kurds (18 men and 15 women), and 35 Ashkenazi men.

Before the work capacity test, the experimental procedures and techniques were explained, and subjects were familiarised with them.   Height and weight measurements were taken and lean body mass was calculated from skinfold thickness measurements obtained using the Harpender caliper. Calf and thigh volumes of the Yemenites and Kurds were measured by water displacement in a cylinder with a thin, mm-graduated side tube. Measurements were taken at three levels - at the malleolus, tibial tuberosity, and the maximum circumference level of the thigh.

The work capacity test was performed on a calibrated (Monark) bicycle ergometer at 50 revolutions/minute.   The subjects were accustomed to the work procedure by a preliminary trial for 5-10 minutes.   Men were tested at 300, 600, 750 and 900 Kpm/min, women at 250, 300, 450 and 600 Kpm/min.   At each work load, subjects were requested to exercise for five to six minutes. During the last two minutes at each exercise level, expired gas was collected. Respiratory gas exchange was estimated from the collected gas samples.   Heart rate was monitored by ECG.   Predicted maximum work capacity was determined by extrapolation of the oxygen uptake to a heart rate of 195/min.

The physical characteristics of men in the four test groups did not reveal significant differences.   However, Yemenite and Kurdish men were heavier and taller than women of the same ethnic origin.   Significant differences in thigh and calf volumes were obtained when the Negev Kurds and Yemenites were compared with their ethnic counterparts from the Jerusalem highland area.   The Negev inhabitants of both ethnic groups were found to have a larger thigh and smaller calf volumes (Table 3).

The predicted values of maximum aerobic work capacity did not produce evidence for either ethnically or seasonally determined differences.   The values obtained from the Bedouins (43.7 ± 6.5 ml/kg BW) and from the Ashkenazi (39.2 ± 5.4 ml/kg BW) were not significantly different from the mean values obtained

Table 3:   Physical characteristics (Negev and Jerusalem)

	Height cm ± S.E.	Weight kg ± S.E.	Calf volume cm^3 ± S.E.	Thigh volume cm^3 ± S.E.
Negev Area				
Kurds				
Males	168.9± 6.75	64.29 ± 6.23	2112± 284	4546± 823
Females	153.9± 3.89	58.65 ±10.85	2224± 518	3696± 174
Yemenites				
Males	164.2± 5.05	64.79 ±10.20	2125± 393	4397± 639
Females	152.6± 6.79	51.07 ±10.50	1696± 424	3466± 862
Jerusalem Area				
Kurds				
Males	166.1± 4.80	68.31 ± 9.90	2802± 353	3789± 533
Females	153.3± 4.15	57.80 ± 7.72	2492± 390	3545± 869
Yemenites				
Males	165.0± 6.67	54.33 ± 7.49	2326± 343	2971± 547
Females	153.1± 4.92	50.50 ± 8.40	2137± 360	2894± 537

Table 4:   Maximum aerobic power (predicted O_2 consumption) at heart rate 195/min

	Ethnic group			
	Kurds		Yemenites	
	Males n=18	Females n=15	Males n=14	Females n=16
	Mean S.D.	Mean S.D.	Mean S.D.	Mean S.D.
Jerusalem Study				
L/min	2.31± 0.42	1.48± 0.20	2.20± 0.22	1.47± 0.27
ml/kg B.W.	34.40± 7.00	25.90± 4.50	39.50± 5.30	29.10± 6.50
	n=19	n=12	n=14	n=16
Negev Study				
L/min	2.89± 0.46	1.63± 0.33	2.98± 0.50	1.77± 0.43
ml/kg B.W.	44.45± 7.05	28.96 ± 8.28	46.91± 5.27	35.40± 7.73

from the Yemenites and Kurdish men. This suggested again that maximum work capacity per unit body weight is not genetically dependant, providing that testing has been carried out by comparable experimental procedures.   Sex-determined differences in work capacity were observed both in summer and in winter examinations of the Yemenites and Kurds, VO_2 max tending to be higher in males than in females. However, this difference was attenuated when work capacity was related to lean body mass, suggesting that difference in fat tissue content of the sexes is the determining factor of the previously described results (Table 4).

An unexpected finding was revealed when work capacity of the same ethnic groups but different geographic location were compared. All the Negev subjects (both Yemenites and Kurds) had higher VO_2 values than their Jerusalem counterparts, as expressed in absolute terms or calculated per kg body weight. For the men, these differences were highly significant ($P < 0.001$); for the women, the values of the Negev residents were higher than that of the Jerusalem group ($\dot{V}O_2$ max l/min, $p < 0.02$; $\dot{V}O_2$ max ml/kg BW, $p < 0.05$).

The relationship between the predicted work capacity and volume of the lower extremities of the Negev and Jerusalem highland residents were tested by regressing $\dot{V}O_2$ max kg BW of all Yemenites and Kurds of both regions against their individual thigh volume/kg BW and their individual thigh/calf volume ratios. Since it was found that the distribution of subcutaneous fat is quite different in men and women, these relationships were treated separately for the males and females of the two ethnic groups.

In males, a definite positive correlation between $\dot{V}O_2$ max kg BW and thigh vol/kg BW was obtained (correlation coefficient 0.46, significant at 0.001 level). An even better correlation was obtained between $\dot{V}O_2$ max/kg BW and the thigh/calf volume ratios (correlation coefficient 0.54), suggesting that thigh volume is more significant in predicting work capacity on a bicycle ergometer (Figures 1 and 2).

Positive relationships between predicted work capacity and body composition have been demonstrated in the past (Davies, 1972; Davies et al, 1973; Elbel, 1949). By multiple lead simultaneous electromyography, Houtz and Fisher (1959) have demonstrated that the muscles of the thigh and calf are equally involved during exercise on a stationary bicycle and therefore it can be assumed that the positive correlation between work capacity and leg volume relate to thigh and calf volumes separately as well. The difference in thigh and calf volumes of the subjects examined in this study made it possible to evaluate separately the relative significance of calf and thigh volume as determining components in the leg volume/work capacity relationships.

As to the work capacity and anthropometric differences observed between the subjects of the same ethnic origin but different geographical location, it can be assumed that they are related to the professional activities of the subjects in

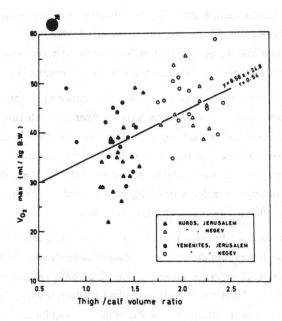

Figure 1:   Environmental, genetic and leg mass influences on energy
expenditure.

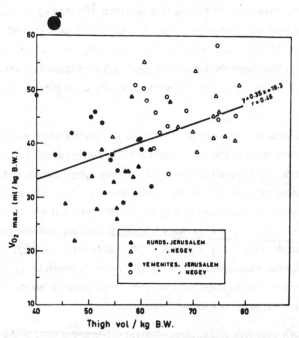

Figure 2:   Environmental, genetic and leg mass influences on energy
expenditure.

both areas. There is a significant difference in the terrain of both locations from where the examined Yemenites and Kurds were recruited. The Negev villages are located on a plain and men living in this area are busy in active farming. The area around Jerusalem is mountainous with man-made terraces. The inhabitants of the Jerusalem highland are engaged mainly in jobs and services in town. Their agricultural activities are sporadic and limited to the orchards on the terraces in the hills. Thus it may be suggested that the occupational activity of the subjects of these two regions is different due to differences in the geophysical conditions of the areas. It seems that habitual occupation affects the anthropometric features of the lower extremities - the Negev inhabitants, both Yemenites and Kurds, exhibited a larger thigh and smaller calf volumes than their counterparts from the highland around Jerusalem.

REFERENCES

Davies, C.T.M. (1972). Maximum aerobic power in relation to body composition in healthy, sedentary adults. Human Biology, **44**, 127.

Davies, C.T.M., Mbelwa, D., Crockford, G. & Weiner, J.S. (1973). Exercise tolerance and body composition of male and female Africans aged 18-30 years. Human Biology, **45**, 31.

Elbel, E.R. (1949). Relationship between leg strength, leg endurance and other body measurements. Journal of Applied Physiology, **2**, 197.

Houtz, S.J. & Fischer, F.J. (1959). An analysis of muscle action and joint excursions during exercise on a stationary bicycle. Journal of Bone and Joint Surgery, **41**, 123.

SOCIOCULTURAL INFLUENCES ON THE WORKING CAPACITY OF ELDERLY NEPALI MEN

C. M. BEALL and M. C. GOLDSTEIN

Department of Anthropology, Case Western Reserve University,
Cleveland, Ohio, U.S.A.

The influence of sociocultural factors on physical working capacity is often ignored or treated as a residual category of little interest. Most catalogues of the factors influencing physical working capacity consider the sociocultural environment as "noise" to control for when analysing what are taken to be "basic" physiological parameters of interest. It is common, therefore, for exercise physiologists to focus on those factors that can directly generate variation in the cardiorespiratory components of physical working capacity, for example on activity or nutritional status, while ignoring sociocultural parameters. Such factors, however, are rarely distributed uniformly and homogeneously in populations, and this distributional variation is determined to a large extent by sociocultural forces. Consequently, an ecological or population perspective on physical work capacity requires consideration of sociocultural factors.

Measuring sociocultural variables, however, is often problematical. Not only are different variables important in different populations (e.g. caste vs. class), but they are often not objectively quantifiable on a universal scale analogous to temperature or haemoglobin concentration. Consequently, determining the relevant sociocultural variables to measure, deciding how to measure them, and then comparing subpopulations or populations, present formidable hurdles to the biological researcher.

A cursory examination of the literature reveals several common ways of dealing with the issue of sociocultural factors, none of which is completely satisfactory. One way has been simply to ignore the possibility of socioculturally generated variation and describe the population under investigation as "peasants", "natives", "residents", etc., i.e. as a category undifferentiated apart from age and sex. This implicitly assumes that variation within such categories is smaller or less interesting than differences between

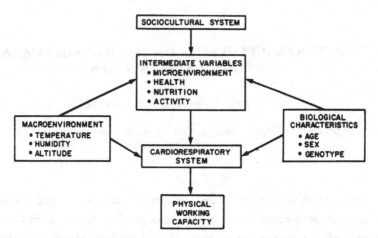

Figure 1:   A general model of influences on physical working capacity
emphasising the way that social factors may generate variation by
acting through a set of intermediate variables.

them and their non-peasant, non-native, etc. counterparts.   A second way has
been to make rudimentary and often ethnocentric subdivisions such as economic
status measured by income.   However, it is clear that in some societies, other
factors such as caste may be more important than class, and this is a matter for
empirical investigation, not *a priori* assumption.   Transposing factors that are
important in western societies to non-western ones implicitly assumes that a
given variable has the same influence in all cultures, and this is, of course, false.
A third way has been to use well-known "ethnic groups", e.g. Tibetans, Hindus.
This implicitly assumes that groups isolated on the basis of a single feature such
as geographical origin or religion are homogeneous in other respects.   These
approaches, therefore, have serious limitations in that they neither pose
precisely framed questions that seek to identify the sociocultural factors
influencing working capacity nor do they determine how such factors operate or
the extent of the effect.

This paper focuses on these issues.   It presents a general model depicting
the paths by which sociocultural influences affect physical working capacity, and
then presents a case study from Nepal.

Figure 1 describes a general model of the way that sociocultural
influences may generate variation in physical working capacity.   The model
depicts the central role of the cardiorespiratory parameters that are the direct
determinants of physical working capacity and through which all other factors

must operate. These cardiorespiratory parameters are subject to a number of unmodifiable and modifiable influences. The unmodifiable influences include (a) direct physical macroenvironmental features, for example low pO_2 availability at high altitudes, and (b) direct biological characteristics, for example the age-associated maximal heart rate.

The modifiable direct influences are grouped together under the label "intermediate variables". This label emphasises that these influences are the link between the sociocultural environment on the one hand and cardiorespiratory factors and physical working capacity on the other, and that these are subject to sociocultural forces. The intermediate variables include microenvironmental variables such as workplace temperature and biological variables such as nutritional status, health and activity patterns. In addition to the sociocultural environment, the intermediate variables are subject to the influence of the physical environmental and biological characteristics, for example a particular physical environment may provide the appropriate conditions for disease vectors and biological ageing processes may cause changes in health status.

Sociocultural influences must act upon these intermediate parameters in order to generate variation in physical work capacity. The small number of physiological mechanisms contrasts with the large number of possible sociocultural influences. For example, the heart rate (HR) response to exercise is determined in part by the level of habitual activity which is influenced by cultural norms for the appropriate behavior of individuals of different age, sex, and social status as well as by variables such as socioeconomic status, occupation and household demography. Consequently, one might hypothesise that the activity, and hence the physical work capacity, of a wealthy grandfather living in a large multigenerational household would differ from that of a poor young widow living alone. There are many possible combinations of factors. Some subdivisions such as the one just described may appear obvious, others may not: for example, poor grandfathers living in large multigenerational households or rich grandfathers living alone. Whether these differ in activity and physical working capacity would be empirical questions for investigation. The insight gained from attention to sociocultural patterns will enable identification of reasonable subdivisions and the development of appropriate testable hypotheses about where and why variation in activity and physical working capacity is observed.

This model is illustrated by data derived from a study examining the relationship between activity, health and physical fitness of the elderly in a non-

Western setting (see Beall et al, 1985a; Beall et al, 1985b).   The study
population are native residents of the rural agrarian community of Chetbesi
(pseudonym), Lamjung District, Nepal.   This is a traditional, rural, low altitude
(918 m) community with a monsoonal climate that is perched on the steep slopes
rising from the Marsyangdi River.   It is located in a roadless rugged mountain
terrain without facilities such as electricity, sanitation or machinery.   Human
and animal muscle power are the sources of energy for transportation and
production.   Farming and herding are the basis of the local economy, and arable
land is the critical economic resource.

Chetbesi is inhabited by several ethnic groups including Nepali-speaking
Hindus who are the subject of this study.   Nepalese Hindu communities are
divided into hereditary immutable social groups called castes into which
individuals are born and remain for their entire lives.   The social system
segments these castes into a broad two-part hierarchy consisting of socially pure
high castes (Brahmins and Chetris) and socially impure, low or untouchable
castes (Kami - blacksmiths, and Damai - tailors and Sarki - leatherworkers).
Associated with these notions of social purity and pollution are sets of
restrictions on behaviour including occupation.   For example, a person born into
the low blacksmith caste cannot adopt tailoring as his occupation, nor could a
tailor adopt smithing.   A person born into the high Brahmin caste cannot plough
fields.   All castes may perform most agricultural tasks, apart from ploughing.
Thus, birth into one of these hereditary social categories greatly dictates the
preferred, permitted and prohibited activities throughout an individual's lifetime.

The Hindu social hierarchy tends to be associated with the economic
hierarchy.   In these communities the members of the two high castes own good
agricultural land and are predominantly middle income or well-to-do in village
terms.   The low castes also own or lease land, usually small plots of poor quality
whose yields suffice for only a few months' food supply.   Therefore they are
forced to rely on providing services and earn their livelihood largely as
craftsmen and manual laborers.   They are typically low income, albeit not
impoverished or destitute.   A few households, in fact, are middle income.

All the native high and low caste Hindu males aged 60+ years, living in
three wards of Chetbesi, were identified by a census survey and invited to
participate in a study of the health of the elderly.   Forty-three (88%) of the
eligible people participated in the study, including 9 Brahmins, 16 Chetris, 6
Damais, 3 Kamis and 9 Sarkis (mean age 67 ± 7, range 60-88).   The physical
working capacity of 34 individuals was measured by a cycle ergometer test using
an intermittent multistage protocol.   Nine men (21%) were excluded from

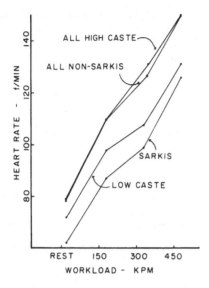

Figure 2: Heart Rate response to three levels of physical work performed by elderly Nepali men: comparison of high vs low caste and of non-Sarkis and Sarkis.

performing the test on medical grounds or by pain or weakness which obviated pedalling. One Sarki failed to give a good effort on the test and was eliminated from analysis. Thus the analysis of physiological data is based on a sample of 33 men with a mean age of 66 ± 5, range 60-77. Measurements of heart rate (HR) and systolic blood pressure were made in the course of the cycle ergometer test. Measurements obtained during the third minute of exercise at the 150, 300 and 450 kpm submaximal workloads are reported here.

A logical initial subdivision for analysis is one comparing high and low caste as this roughly coincides with a culturally important distinction and with socioeconomic status. Figure 2 demonstrates that there is significant physiological variation in response to cycle ergometer exercise; low caste men have lower HR at 300 and 450 kpm effort. Figure 2 demonstrates also that the coarse subdivision into high and low caste is misleading and confounded. A single low caste, the Sarkis, accounts for the contrast.

The Sarkis differ systematically in cardiorespiratory characteristics measured before, during and after exercise. They have lower resting HR, lower HR at 150, 300 and 450 kpm submaximal workloads and lower HR after terminating exercise (see Beall et al, 1985a, for details). This indicates a more efficient, well-trained physiological adaptation to exercise stress. The

Table 1:   Comparison of selected health and nutritional characteristics of
Chetbesi Sarkis and non-Sarkis

	Sarkis			Non-Sarkis		
	$\bar{X}$	S.D.	N	$\bar{X}$	S.D.	N
Health status:						
FVC (dl BTPS)	261	59	6	268	79	23
FEV-1 (dl BTPS)	187	42	6	186	63	23
Nutritional status:						
BMI (kg/cm^2)	18.4	1.1	8	17.6	2.1	25
% Body fat*	14.4	4.1	8	13.8	6.0	25

*Calculated after Durnin and Wormersley (1974), Table 5 equation for triceps,
subscapular and suprailiac skinfolds.   Males aged 50+.

close correspondence of the lines depicting the high castes and those depicting all the low caste non-Sarkis emphasises the physiological similarity of the Damais, Kamis, Brahmins and Chetris, and highlights the contrast between these four groups and the Sarkis.   The non-Sarkis and Sarkis do not differ in HR and systolic blood pressure at peak tolerated work effort.

The model in Figure 1 suggests a systematic approach to developing an explanation for this intracultural variation in physical work capacity.   The macroenvironment, assessed by temperature, humidity and altitude, in which the Sarkis and non-Sarkis live and work does not differ as they are part of a single residential community.   The unmodifiable biological characteristics of Sarkis and non-Sarkis do not differ as they are all native males of the same average age and age range.   Turning to the intermediate socioculturally modifiable variables, most activities are conducted out of doors and therefore the microenvironments do not differ.   The health status does not differ as both groups were subject to the same health screening to determine eligibility for participating in the cycle ergometer test.   Other assessments of health status such as measurements of pulmonary function by FVC and FEV 1.0 provide further evidence of similarity (Table 1).   Nor does the nutritional status of Sarkis differ from that of non-Sarkis.   Anthropometric assessments of nutritional status including body mass index, and the percent of body fat estimated from skinfolds do not differ (Table 1).

The similarity of the macroenvironment, of the unmodifiable biological characteristics, of the microenvironment, and of the health and nutritional status of the Sarkis and the non-Sarkis suggests that the fitness difference is most likely the outcome of differences in habitual activity patterns. Performance of heavy tasks causing HR elevation is probably the source of training effects influencing physical working capacity.

For complex social and historical reasons, the Sarki caste in this area has formed the agricultural labour force for the high castes, while the Kami and Damai castes have subsisted mainly by practising their less strenuous traditional crafts of smithing and tailoring. The Sarkis perform heavy work for the Brahmins and Chetris, particularly agricultural labour. They also work as porters, masons and construction workers. Consequently, the Sarkis' subsistence mode is distinct from both the high castes and the other low castes in that it entails perfoming hard labour throughout the year.

Furthermore, Sarkis are likely to continue to perform heavy manual labour throughout their lives. Eight of the nine Sarkis in this sample of men, 60-77 years old, perform manual labour for others: ploughing/levelling fields, stone cutting, masonry, construction and portering. The sole exception is a 77-year-old who had discontinued ploughing with the current season. In contrast, just three of the nine other low-caste elderly men hire out occasionally as labourers in addition to plying their traditional trades which entail moderate and low activity. The Brahmins and Chetris do not supplement their agricultural incomes with manual labour for others although they may perform heavy tasks for themselves.

Direct observation of the daily activities of Sarkis and non-Sarkis confirm these self-reports. While the Sarkis performed heavy labour such as ploughing or load-carrying on half (5/10) of the days of direct observation, non-Sarkis performed heavy labour on just one quarter (22/87) of the days of observation. Moreover, when undertaking heavy tasks such as ploughing, the Sarkis work at them for long periods: a median of 58% of the observed day (12-13 hours of observation) compared with 16% for non-Sarkis. Heavy tasks occupied 47-66% of the day on 4 of the 5 days when Sarkis engaged in such tasks, whereas they occupied an equivalently large proportion in just 5 of the 22 days when non-Sarkis engaged in them. Both the frequency and the duration of heavy work differ between Sarkis and non-Sarkis in a way that could produce the measured differences in physical working capacity. Walking in the rugged terrain has a component of heavy activity, but Sarkis and non-Sarkis walk outside their house compounds for equivalent amounts of time. The Sarkis spend a smaller proportion of the day engaged in light activity. This is additional evidence that occupational activity requirements are the source of differences in fitness.

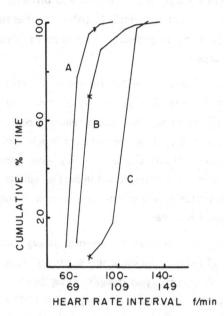

Figure 3:   Cumulative frequency distribution of day-long HR of three Nepali men. A = 60-year-old Sarki; B = 70-year-old non-Sarki; C = 70-year-old Sarki.   The x on each curve denotes 50% of the age-predicted maximal heart rate. (Redrawn from Beall et al, 1985a.)

Overall, Sarkis spend more time engaged in heavy labour, the same amount of time walking and less time at light tasks.

Day-long HR monitoring confirms that Sarkis' daily workload is generally heavier.   The day-long average HR of Sarkis (78 ± 16, n = 5) and non-Sarkis (83 ± 11, n = 28) do not differ.   This similarity is evidence that the more fit Sarkis do more work.   Because of their greater fitness and their lower resting HR, a heavier workload is required to elevate Sarkis' HR to the same level.   If the workloads were the same, the well-trained Sarkis would have lower HR.

This is illustrated in Figure 3 by cumulative frequency curves of day-long HR for a 60-year-old Sarki (A) and a 70-year-old non-Sarki (B) performing the same activities.   The Sarki spent 81% of the day engaged in light activity, and 19% walking, and the non-Sarki spent 80% and 20% respectively.   The Sarki's curve lies to the left of the non-Sarki's and illustrates his performance of the same type of activity at a lower HR. The Sarki's day-long average HR is 41% of his age-predicted maximum HR, while the non-Sarki's day-long average HR is 53% of his age-predicted maximum HR.   The cross on each curve denotes the

HR that is 50% of the age-predicted maximum HR. The cross is much lower on the non-Sarki curve, indicating a larger proportion of HR above this value. Thus the non-Sarki's workload is more intense relative to his age-predicted maximum, although the two were performing similar activities.

The curve marked C in Figure 3 demonstrates that ploughing and levelling fields, a task often performed by Sarkis, constitutes a very high relative workload. The cumulative HR frequency curve for this 70-year-old Sarki who spent 66% of the observed day intermittently ploughing and levelling fields, lies far to the right of the other two curves. Moreover, most of the day's activities occurred above 50% of the age-predicted maximum HR. His day-long average HR is 71% of his age-predicted maximum HR. The HR data are additional evidence that spectific tasks most frequently performed by Sarkis are a great physiological strain.

This example illustrates the power of sociocultural variables to generate variation in physical working capacity in old age. This variation is accounted for by variation in occupational activity levels. In turn, occupation is a function of birth into a socially defined group, a caste. The Sarkis maintain a physical activity pattern of frequent and long periods of strenuous manual labour, even in old age, and have a greater physical work capacity in old age. The magnitude of the effect is substantial. For example, in this population, the regression of submaximal HR on age reveals HR differences of 7 f/minute/decade at 150 kpm. Working at 150 kpm, Sarkis' HR are 13 f/minute lower than non-Sarkis of the same age, roughly equivalent to a two-decade age difference.

So far this example illustrates the intricacies of accurate identification and definition of important subgroups for epidemiological analyses. There is a further level of analysis to consider: why do Sarkis maintain high occupational activity levels as they grow old? After all, a widely commented upon aspect of Hindu culture is that elderly parents should depend upon their sons for security in old age. Since all but two of the sample (both non-Sarkis) are living with sons, it appears that the opportunity for "retirement" exists. That Sarkis and others do not retire, and that Sarkis maintain high levels of activity, is explicable by understanding the sociocultural dynamics of households.

The reasons underlying this activity difference relate ultimately to access to and control over arable land, the basic economic resource in this rural area. Land is held in the father's name and actually controlled by him until his death. The culturally ideal role for an elderly male in Nepalese Hindu society is that of head of an extended household which includes his spouse, married sons and their families, and unmarried children.   All household members are under the authority of the household head and are expected to participate in the household economy either by working the family land under his direction or by contributing wages to him. High caste males expect to give up most physically arduous tasks as they grow old but they expect to retain managerial control over their agricultural land and their household economies. By relinquishing the physically onerous tasks to their sons, they become dependent upon them to perform these tasks, but it is a dependence similar to that of a master on a servant.   In this case the son is willing to do the work because of the land he will inherit eventually and use to establish his own independent household.

Low caste males face a different set of circumstances as they age. They are generally without a large estate of arable land and they expect to work for wages.   The wealth of the family is the sum of the collective efforts of the members, each of whom contributes most of his daily wages to the household head, the father, who also works daily.   Low caste elderly males have only their own wages (i.e. no or little land) to contribute, and therefore feel they must continue to work and earn in old age if they are to maintain their household position and status.   Elderly low caste males consider that once they cease working and earning they quickly lose managerial control and authority in the household and become dependent upon a son who, as the main income earner, assumes the role of household manager.

This dependence is completely different from that of the high caste elderly, for it is the dependence on the penniless upon the provider.   Therefore low caste elderly males strive to avoid becoming completely dependent on their sons in this culturally pejorative sense by continuing to work and earn money as long as they possibly can.   Through work, not only can they contribute to meeting the household's expenses, but they also generate money which they control and can spend as they please.   These powerful social forces pressure elderly low castes into continuing to earn wages for as long as possible.

Thus the experience of growing old is fundamentally different for low and high caste males. All low caste are compelled by real subsistence needs and by considerations of self-esteem to continue to work for income until absolutely unable to do so physically.   For Sarkis this means maintaining high levels of activity for as long as possible, and one unintended consequence of this is greater working capacity.   In this case the intermediate variable linking intrapopulation variation in physical work capacity to sociocultural influences is physical activity, and the influences themselves are membership in a hereditary social group and elderly men's desire to maintain family authority and self-esteem.

The above findings raise the broader question as to whether each unique sociocultural system generates a unique pattern of variation in physical working capacity that is uniquely explicable, or whether the more desirable scientific goal of the development of general models of relationships is possible.   The Chetbesi data suggest the latter.   For example, it could be hypothesised that traditional peasant societies with hierarchical and stratified social structures generally contain one or more subgroups which perform the bulk of the hard work and have a high physical working capacity.   In this particular case, a highly stratified peasant society, the hard-working group, comprises the full adult age range of one stratum - the untouchable Sarkis.   In a nonstratified society, however, it could be hypothesised that the pattern is either (a) completely different in that the hard-working group includes virtually everyone in a narrow age range, or (b) that it is similar in that a non-hereditary landless (or poor) segment of the population comprises the hard-working group.   We suggest that there may in fact be only just a few sociocultural patterns of allocating hard work and therefore only a few general patterns of intracultural variation in working capacity.

Although sociocultural factors are more often than not either ignored or treated cursorily, this study suggests that they play a central role in understanding the distribution of physical work capacity in human populations. Because they affect the distribution of the intermediate variables that affect the key cardiorespiratory parameters, sociocultural factors are essential for explaining how and why differences in physical work capacity exist within a population.   However, as indicated in this study, in a culturally alien setting it may be difficult to ascertain what sociocultural factors are pertinent and how they may be held constant.   In such unfamiliar contexts, familiar cues may not be present or, if present, may not be applicable.   For this reason, it is important to include a serious sociocultural component when investigating variation in physical work capacity, whether the goal is to obtain explanations at the level of human physiology or at the level of human population differences.

ACKNOWLEDGEMENT

This research was supported by grants from the National Science Foundation (BNS 82-19188), National Geographic Society (2641-83) and the American Federation for Aging Research.

REFERENCES

Beall, C. M., Goldstein, M. C. & Feldman, E. S. (1985a).   Social structure and intracohort variation in physical fitness among elderly males in a traditional third world society.   Journal of the American Geriatrics Society, **33**, 406-412.

Beall, C. M., Goldstein & M. C. Feldman, E. S. (1985b).   The physical fitness of elderly Nepalese farmers residing in rugged mountain and flat terrain. Journal of Gerontology,   **40**, 529-535.

Berger, E. (1982).   Applied Exercise Physiology.   Philadelphia: Lea & Febiger.

Durnin, J.V.G.A. & Womersley, J. (1974).   Body fat assessed from total body density and its estimation from skinfold thickness: measurements on 481 men and women aged from 16 to 72 years.   British Journal of Nutrition, **32**, 77-97.

ENERGY EXPENDITURE AND ENDEMIC DISEASE

MALARIA: WORKING CAPACITY IN A HOLO- AND MESOENDEMIC REGION OF LIBERIA

E. BENGTSSON, P. O. PEHRSON, A. BJORKMAN, J. BROHULT,
L. JORFELDT, P. LUNDBERGH, L. ROMBO, M. WILLCOX and A. HANSON

*Department of Infectious Diseases, Karolinska Institute,
Roslagstulls Hospital, Stockholm, Sweden*

INTRODUCTION

There is a widespread but preconceived opinion held by many laymen, medical personnel, politicians or aid organisations, that working capacity is impaired in chronic parasitic diseases. There are clearcut reasons to suggest this when, for example, Schistosomiasis is followed by liver fibrosis or malignancy of the urogenital tract, or when malaria has been complicated by perimyocarditis. Generally, however, we know little about the consequences of toxins or other antigens spread systemically without giving rise to permanent morphological changes in various organs. One of the examples would be malaria in which, more or less continuously, liberation of antigens foreign to human proteins occurs. Though not demonstrable in histopathology, this may nevertheless impair the function of visceral tissues.

Many years ago, Bentgsson (1956, 1957) reported that physical working capacity was impaired in acute bacterial and viral infections in patients who had demonstrable involvement of the cardiovascular or respiratory systems, as well as in cases of anaemia, according to the aetiology and severity of the disease. There are few investigations of working capacity in chronic infectious and parasitic diseases which do not involve respiratory, cardiac or blood tissues, although there are known to be important practical implications. The reason for the paucity of scientific anlayses may be the difficulty in finding adequate and reliable methods. Of the socioeconomic implications, the level of wages, living standards, absenteeism from work, etc., have been examined. Among the earliest of such analyses, Fenwick and Figenschou (1972) studied the effect of Schistosoma mansoni infection on the productivity of cane-cutters on a sugar estate. In other investigations adequate physiological methodology has been applied mostly in studies of anaemia.

PARAMETERS OF WORKING CAPACITY

In this study, search was made for different parameters of muscle force and endurance in physical exercise, and the most adequate test was found to be that using the Sjostrand bicycle ergometer (Sjostrand, 1967). The principle of this method is that the work capacity is related to the total mass of muscle tissue and at the same time also to factors such as age and body weight. There is a direct interrelationship between work and heart volume, oxygen consumption and in particular the total haemoglobin. Athletes thus have a working capacity above the normal limits; strict bed rest will restrict it (Figure 1). Patients with latent heart insufficiency, e.g. in perimyocarditis, show a deviation from the normal relationship between working capacity and heart volume as indicated by a decreased stroke volume and cardiac dilatation (Table 1). Similarly, abnormalities in respiratory organs and blood must be taken into account in the determination of working capacity.

In the development of the bicycle ergometer test it was shown that almost all healthy volunteers and convalescents were able to work on the bicycle with a steady state heart rate of up to 170 beats per minute which represents a high submaximal effort. This is defined as the working capacity (W_{170}). The W_{170} as measured in kilogram-metres (Figure 2) increased with age (5 to 40 years). In adult males the W_{170} was 1200 kgm which is about 300 kgm more than in adult females (900 kgm), which in turn was the W_{170} of male adolescents – and also just the same figure was found in adult male Africans from a holoendemic region of malaria. Figure 2 also demonstrates the importance of training: both female and male gymnasts had a much higher performance.

Besides the total muscle mass there are three main factors limiting the W_{170}: respiratory function, the stroke volume of the heart and the oxygen transportation of the blood. The particular parameters applied in the following study were the levels of haematocrit and total haemoglobin.

Material and Results

The aim of the study was to find out if any factors other than body dimensions were limiting the W_{170} of adults in a highly endemic malarial region. In north-east Liberia we had the opportunity to compare adult males in holoendemic areas (34 farmers; parasitic rate in children 95%) with males in meso- and hypoendemic towns (49 workers at the LAMCO Mining Co. – 15 policemen and 35 staff members of the company; parasitic rate in children about 10%) (Hedman et al, 1979; Brohult et al, 1981). Also, we made a comparison with a group of 167 Swedish males. The parasite rate in the farmers

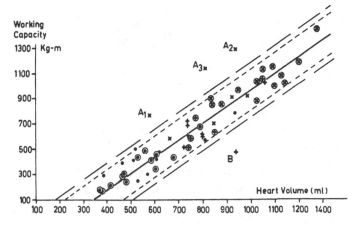

Figure 1: W_{170} in relation to heart volume in healthy volunteers.   A1, 2 and 3 relate to athletes, B to one untrained person.

Table 1:   Working capacity, body weight and heart volume in 19 patients with ECG abnormalities suggestive of myocarditis.

	Working capacity in relation to body weight	Working capacity in relation to body weight	Heart volume in relation to body weight
Abnormal	7	12	2
Borderline	0	4	3
Normal	12	3	14
Total no. of observations	19	19	19

was 31% and in the mesoendemic group 6%.   The demographic data (Table 2) shows that height, weight and absolute value of blood volume were all significantly higher among the policemen than among the workers.   When, however, related to the body weight, the blood volume was the same in the two groups. The haemoglobin concentration was higher among the workers compared with the staff members.

No difference was observed in total Hb between holo- and mesoendemic groups, but the haemoglobin concentration was lower in the holoendemic farmers who, by way of compensation, had a greater plasma volume.   The difference in this respect may be due to the heavier load of parasites in the rural than in the urban area.   Subjects with positive and negative malaria smears did not differ

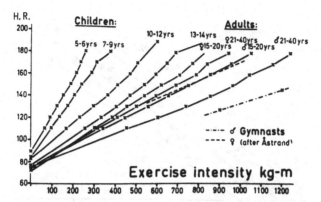

Figure 2:   Exercise intensity in relation to heart rate in Swedish children and young adults.

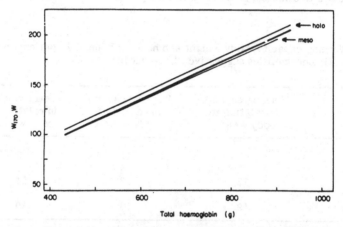

Figure 3:   The relationship between working capacity and the amount of total haemoglobin in the holo-endemic group (farmers) and the meso-endemic group. There is no significant difference between the regressions for the holo- and meso-endemic groups.

significantly with regard to haemoglobin concentration, total haemoglobin and W_{170} (Table 2).

W_{170} was about 150 watts (corresponding to 900 kgm) in all groups. If there was any difference at all between the groups it was to the advantage of the farmers from the holoendemic area who had 8% higher W_{170} than the values from the mesoendemic volunteers. This could be related to the larger amount of total Hb in the farmer group, although the difference was not significant. Nor

Table 2: Age, height, weight, haemoglobin concentration, haematocrit, plasma volume, blood volume, total haemoglobin and physical working capacity (W_{170}) in 39 farmers living in holo-endemic areas and 99 men living in meso-endemic areas (workers, policemen and Lamco staff). P shows the level of significance between the holo-endemic and meso-endemic groups.

Variable no.	Holo-endemic group						Meso-endemic group				P
	Farmers (n=39)		Workers (n=49)		Policemen (n=15)		Staff (n=35)		W + P + S (n=99)		
	Mean	S.D.	Mean	S.D.	Mean	S.D.	Mean	S.D.	Mean	S.D.	
Age (years)	24.7	6.7	29.8	5.3	28.5	3.5	28.4	7.7	29.1	6.1	0.001
Height (m)	1.70	0.06	1.67	0.06	1.74	0.07	1.70	0.06	1.69	0.07	NS
Weight (kg)	1.4	8.0	61.0	8.9	68.7	6.4	65.1	11.0	63.6	9.7	NS
Hb (gl^{-1})	148	15	156	11	151	14	149	12	153	12	0.05
HCT (%)	45.8	4.3	46.8	3.1	46.3	4.0	46.2	3.1	46.5	3.2	NS
PV (1)	2.99	0.43	2.61	0.39	2.78	0.42	2.80	0.36	2.70	0.39	0.001
BV (1)	5.12	0.59	4.54	0.60	5.16	0.55	4.83	0.56	4.74	0.61	0.01
Total Hb (g)	689	104	646	89	704	89	652	90	657	91	NS
W_{170}, W	159	37	148	32	149	27	141	29	146	30	0.05

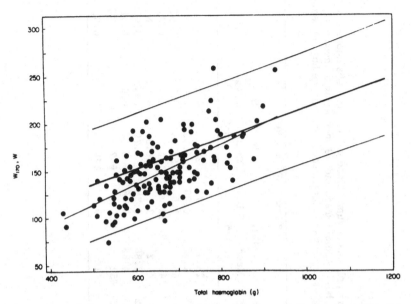

Figure 4:   The relationship between working capacity and the amount of total
haemoglobin in Liberian males as compared to Swedish normal
material.   The linear regression line (heavy), and the limits of the
95% confidence interval for the Swedish material, and the linear
regression line (thin) for the Liberian material, are indicated.

did the regression lines of holo- and mesoendemic areas differ (Figure 3).   The
coefficient of correlation was about the same (r = 0.60) in all groups, the Swedish
material included.

Hookworm was observed in 28% of those in the holoendemic region and in
only 1% in the mesoendemic area.   Onchocerciasis was observed in 44% and 1%
respectively.   In neither case did the subjects differ significantly as regards
W_{170}, Hb concentration or total Hb.

Finally, comparison was made with 167 Swedish males of similar age by
plotting the values of W_{170} and total Hb of the Liberians on the regression line
of the Swedish material (Figure 4).   All or almost all were distributed within the
95% confidence interval.

On the assumption that a mean parasite rate of 6% is indicative of
chronic malaria in the adult population, that regular weekly chloroquine prophylaxis should eliminate the malaria infection, and that this eradication might
improve the physical work performance, the question of whether regular

Table 3:   Parameters of work capacity in 17 subjects on prophylaxis and 15 controls on 2 occasions, 1 year apart

	HB g/l^{-1}	HCT %	Blood volume (litres)	Total Hb(g)	W_{170}	+ve Smears
Subjects on Prophylaxis						
Observation						
1	155 ± 12	46.8± 2.8	4.64± 0.59	657 ± 100	746 ± 158	1
2	155 ± 10	47.5± 2.9	4.63± 0.59	653 ± 78	737 ± 155	
Controls						
Observation						
1	153 ± 15	46.4± 3.4	4.36± 0.37	607 ± 74	648 ± 148	5
2	151 ± 8	45.3± 2.7	4.41± 0.41	606 ± 65	639 ± 168	

chloroquine prophylaxis might improve the working capacity was investigated (Pehrson et al, 1984). This investigation included 17 subjects on a 300 mg base chloroquine weekly prophylaxis, and 15 controls in an open study. They were all examined regularly and were told to call for examination whenever they had any signs of acute disease and not to take any drugs without consulting the research officer. The groups were followed for one year (Table 3). No significant change was found in any of the parameters although the values of hematocrit increased in the prophylaxis group and decreased in the control group. However, total Hb and W_{170} did not change in any of the groups.

CONCLUSIONS

These results imply that differences in working capacity between groups of Liberians and between Liberians and Swedes usually refer to body dimensions. In absolute terms W_{170} of the Liberian males was about 30% lower than in Swedish males and of the same level as Swedish male adolescents and Swedish female gymnasts. The training factor in the different groups might be impossible to estimate or compare. Thus, when correlated with the total Hb, the relative value of W_{170} was similar in all groups: bush farmers in holoendemic areas, policemen and staff members in urban mesoendemic areas and Swedes in non-endemic regions. Nor were there any differences between people with positive and negative parasite smears. Regular malaria prophylaxis for workers in urban mesoendemic areas did not seem to influence physical performance.

REFERENCES

Bengtsson, E. (1956).   The working capacity in normal children, evaluated by submaximal exercise on the bicycle ergometer and compared with adults. Acta medica Scandinavica **154**, 91.

Bengtsson, E. (1957).   Working capacity and heart volume in patients with electro-cardiographic abnormalities suggestive of acute myocarditis. Acta medica Scandinavica **159**, 499.

Brohult, J., Jorfeldt, L., Rombo, L., Bjorkman, A., Pehrson, P.O., Sirleaf, J. & Bengtsson, E. (1981).   The working capacity of Liberian males: a comparison between urban and rural populations in relation to malaria. Annals of Tropical Medicine and Hygiene, **75**, 487.

Fenwick, A. & Figenschou, B.H. (1972).   The effect of *Schistosoma mansoni* infection on the productivity of cane cutters on a sugar estate in Tanzania. Bulletin of the World Health Organisation, **47**, 567.

Hedman, P., Brohult, J., Forslund, J., Sirleaf, J. & Bengtsson, E. (1979).   A pocket of controlled malaria in a holo-endemic region of West Africa. Annals of Trpical Medicine and Hygiene, **73**, 317.

Pehrson, P.O., Bjorkman, A., Brohult, J., Jorfeldt, L., Lundbergh, P., Rombo, L., Willcox, M. & Bengtsson, E. (1984).   Is the working capacity of Liberian industrial workers increased by regular malaria prophylaxis?   Annals of Tropical Medicine and Hygiene, **78**, 453.

Sjostrand, T. (1967).   Clinical Physiology.   Scandinavian University Books. Stockholm: Svenska Bokforlaget.

SCHISTOSOMIASIS: FIELD STUDIES OF ENERGY EXPENDITURE IN AGRICULTURAL WORKERS IN THE SUDAN

K. J. COLLINS[1], T. A. ABDEL-RAHAMAN[2] and M. A. AWAD EL KARIM[3]

[1] Medical Research Council, London School of Hygiene & Tropical Medicine, London, U.K.

[2] Department of Physiology, University of Juba, Juba, The Sudan.

[3] Department of Community Medicine, University of Kuwait, Kuwait.

INTRODUCTION

The fatigue and debility commonly associated with schistosomiasis lend support to the popular belief that schistosomiasis diminishes physical working capacity and productive work output in infected populations (Gelfand, 1963; Forsyth & Bradley, 1966; World Health Organisation, 1967; Cheng, 1971). More objective studies on the nature of the incapacity to work and the extent to which the disease influences physiological performance have been conflicting. In controlled work tests on East African (Davies, 1972) and Bantu (Walker et al, 1972) schoolchildren with *Schistosoma mansoni* infection, the physiological response to exercise was found to be unimpaired. Assessment of physical performance and lung function in young adult males in a laboratory study, on the other hand, suggested that some physiological parameters of physical fitness were affected by the disease and that these were significantly improved after treatment (Omer & El Din Ahmed, 1974).

In our early studies of work performance in cane cutters in an endemic area of schistosomiasis in the Sudan, no significant differences were found between infected and non-infected groups in laboratory tests of submaximal and predicted maximal aerobic power (Collins et al, 1976). These studies were designed to account for other sources of variability between the groups such as differences in stature, nutrition and intercurrent malaria or hookworm infection. A more important source of bias, however, appeared to be in the selection for investigation of a population of cane cutters who were relatively fit and from whom the more seriously ill and heavily infected were excluded in the screening procedure. In a subsequent study (Awad El Karim et al, 1980) on an indigenous population of farmers in the Gezira area of the Sudan a statistically significant difference in $\dot{V}O_2$ max amounting to 18% was found between non-infected and highly infected agricultural workers.   This difference was not due to factors of age, body weight, stature, lean body mass, leg muscle volume or nutritional status.   The effects of schistosomiasis on working capacity could, however, be shown to be related to the intensity of infection in terms of the number of schistosome eggs excreted, and this in turn was correlated with the haemoglobin concentration of individuals (Awad El Karim et al, 1980).

Subsequently it was possible to investigate the effect of treatment with schistosomicidal agents on farmers who were living in the endemic area.   It was found that after a period of up to a year following chemotherapy during which time no re-infection had occurred, the $\dot{V}O_2$ max of agricultural workers improved in standard laboratory tests by up to 20% (Awad El Karim et al, 1981).

Although a significant correlation has been found between an individual's work capacity in laboratory tests and actual energy expenditure in a sustained work situation (Collins, 1983) there is little direct evidence of the deleterious effects of schistosomiasis on self-paced work in the working environment.   A field study was therefore planned to assess Sudanese agricultural workers undertaking normal working tasks using the Oxylog portable system for measuring ventilation rate and oxygen consumption (Humphrey & Wolff, 1977).   The objective was to confirm the effects of S. mansoni and its treatment observed previously in extensive laboratory tests and to extend the observations to the practical work situation.   These studies were undertaken in 1981/82.

THE OXYLOG IN FIELD TRIALS

The use of Douglas bags or Kofranyi-Michaelis respirometers under field conditions often introduce inaccuracies in measuring energy expenditure because of the bulk of the apparatus and because gas samples may change in composition with selective loss of CO_2 if gas samples have to be transported any distance (Shephard, 1955).   The development of portable and reasonably lightweight electronic respirometers capable of direct measurement of oxygen consumption and ventilation have greatly facilitated these measurements in the field.   The Oxylog is one such instrument designed to analyse expired air in the range of 6 to 80 $1\,min^{-1}$ ventilation and 0.25 to 3.0 $1\,min^{-1}$ oxygen consumption.   It is therefore suitable for measuring oxygen consumption during normal work and submaximal exercise.

In the present field studies validation of the Oxylog against other standard methods was carried out beforehand in the Faculty of Medicine, University of

Khartoum. Oxylog measurements were made on 14 young male subjects immediately after using a KM respirometer, and secondly while collecting expired air in a Douglas bag during bicycle ergometer exercise (Abdel-Rahaman, 1984). The experimental arrangement in the ergometer test was such that expired air passed first into the Oxylog and then into the Douglas bag in series. In the KM comparison expired air was analysed by a conventional Haldane apparatus and in the ergometer test by a Servomex O_2 analyser. Stepping tests were made in the first experiment with the KM using a 30 cm platform at 12, 18 and 24 steps min^{-1} for five minutes at each work rate, and on the ergometer for five minutes at each of three work rates (300, 600 and 750 kpm min^{-1}).

Table 1 shows the comparisons between Oxylog measurements and the other standard methods. An analysis of variance demonstrated that there was no significant difference between the Oxylog and other methods over the range of submaximal work rates expected to be encountered in the subsequent field trials. Other laboratory evaluations also confirm that the Oxylog is sufficiently accurate for the reliable determination of oxygen intake in the field (Belyavin, Brown & Harrison, 1981; Dounis, Steventon & Wilson, 1980).

During tasks such as digging, the Oxylog was well-tolerated by all workers and little practice was required to accustom the men to wearing the face mask while working. Leaks around the mask (a modified RAF face mask with inflatable rim) were experienced only occasionally and were readily identified and corrected.

ENERGY EXPENDITURE IN AGRICULTURAL WORKERS IN THE GEZIRA

Field work was undertaken in the Gezira, a flat, fertile region in the Sudan which lies to the south of Khartoum in the triangle formed by the Blue and the White Niles before they join at Khartoum. The region occupies about 5.2 x 10^6 acres of which about 2.1 x 10^6 acres are under irrigation. The Gezira villages of Dulga and GadElein were selected for the study, each lying about 120 Km south of Khartoum and 2 Km apart. Each village has a population of about 1000. The major endemic diseases, as in other parts of the Gezira, are schistosomiasis, malaria and other diarrhoeal diseases.

Screening procedures undertaken as part of the ongoing 'Blue Nile Health Project' facilitated the selection of subjects according to age, occupation and *S. mansoni* egg loads. Blood films were examined for malaria parasites and if positive these subjects were excluded from the tests, as were those suffering from hookworm (the incidence of hookworm is low in the Gezira population), asthma, hypertension, heart disease or diabetes. In addition to the para-

Table 1:   Comparison of Oxylog with other standard methods for measuring
ventilation and oxygen consumption in exercise tests.

Work rate	Ventilation rate (mean ± S.D.)	
(a) Steps/min (n=8)	Oxylog (L.min^{-1})	K.M. (L.min^{-1})
12	24.59 ±3.49	25.03 ±3.6
18	33.59 ±7.07	34.26 ±7.04
24	40.96 ±7.98	41.52 ±7.79
ALL (n = 24)	33.03 ±9.24	33.60 ±9.22
(b) Kpm.min^{-1} (n=6)	Oxylog (L.min^{-1})	Gas meter (L.min^{-1}
300	24.53 ±2.85	23.95 ±2.80
600	32.02 ±3.81	31.22 ±3.74
750	40.35 ±1.73	39.35 ±1.91
ALL (n = 18)	32.31 ±7.19	31.51 ±7.03

Work rate	Oxygen consumption (mean ± S.D.)	
(a) Steps/min (n=8)	Oxylog (L.min^{-1})	Haldane (L.min^{-1})
12	0.88 ±0.15	0.88 ±0.15
18	1.25 ±0.21	1.25 ±0.20
24	1.68 ±0.33	1.68 ±0.32
ALL (n=24)	1.27 ±0.41	1.27 ±0.40
(b) Kpm.min^{-1} (n=6)	Oxylog (L.min^{-1})	Servomex (L.min^{-1})
300	1.04 ±0.14	1.03 ±0.14
600	1.52 ±0.29	1.50 ±0.28
750	1.98 ±0.22	1.98 ±0.24
ALL (n=18)	1.51 ±0.44	1.50 ±0.45

sitological screening procedure, each subject completed a questionnaire concerning previous medical history, social status and work habits. Care was taken to ascertain the correct age of the subjects which was checked by information relating to known local or family events. A clinical examination, anthropometric measurements, pulmonary function tests and a blood sample for haematological and biochemical investigations, were taken before the subjects participated in the work study.

The subjects from Gezira villages included 187 males (mean age 32 ± 11 years) of whom 55 were free of *S. mansoni* and the others infected, and 48 canal cleaners, most of whom were heavily infected with *S. mansoni*. The canal cleaners lived in small villages in the same area of the Gezira. The intensity of *S. mansoni* infection was assessed according to the number of eggs excreted in stools (/g faeces) and the subjects were divided into three groups - control (0 eggs), light/moderate (1-999 eggs) and heavy (>1000 eggs).

During the work study, each subject was tested at his usual site of work early in the morning (usually at 6 am.) and before breakfast so that the hot hours of the day were avoided. The main tasks performed were digging or shovelling, the most tiring tasks in agricultural work which are performed during most of the year, and canal cleaners performed weed-cutting tasks by the side of the irrigation canals. After becoming accustomed to wearing the mask of the Oxylog each subject rested, standing, for 5 minutes, then worked at a consistent natural pace, digging, shovelling etc. for 5 minutes, and then rested again for a further 5 minutes.

In Table 2 results are presented of the field study determinations of energy expenditure in 235 farmers and canal cleaners studied in 1981/82. In the heavily infected farmers there was a reduction in energy expenditure for the same task compared with control or lightly infected groups. This was statistically significant only in the initial resting period and could be ascribed to the significantly smaller body weight of the heavily infected farmers. In the canal cleaners group, however, the heavily infected showed a significantly lower $\dot{V}O_2$ during work ($p < 0.05$) with no significant difference in lean body mass or stature. Overall, the canal cleaners performed at a much lower $\dot{V}O_2$ than the farmers. The task of canal cleaning and cutting foliage by the canal bank is a less physically-demanding task than digging for 5 minutes performed by the farmers. There were, however, lower resting levels of $\dot{V}O_2$ in the canal cleaners and also lower haemoglobin levels.

The results of a previous laboratory study of aerobic capacity by ergometer tests are presented in Table 3 for another group of farmers and canal cleaners from the same area of the Gezira studied in 1978/79. Here the work task on the bicycle ergometer was more standardised and the physique of those taking part was similar in all groups. Again there was a decrease in the work capacity of the heavily infected canal cleaners' group and this was accompanied by a significant decrease in haemoglobin concentration. Cotes (see this volume) argues that the power to weight ratio ($\dot{V}O_2$ max Kg^{-1} body mass) is an invalid index because of a significant constant term in the regression relationship between $\dot{V}O_2$ max and body mass. The power output at the mean mass of the group or per unit of body muscle (or body cell mass) is seen as a more acceptable alternative. Predicted $\dot{V}O_2$ max for the groups based on mean lean body mass gives a similar result to that reported in Table 3.

Table 2: Oxylog measurements of energy expenditure in field work in Gezira populations infected with S. mansoni (235 males studied in 1981/82). Means ± S.D.

	Farmers			Canal Cleaners		
	Control	Light-moderate infection	Heavy infection	Control	Light-moderate infection	Heavy infection
n	56	114	17	7	23	18
Age (years)	32 ± 9	32 ± 11	26 ± 10*	38 ± 7	37 ± 6	35 ± 9
Height (cm)	168 ± 8	167 ± 7	169 ± 8	174 ± 5	169 ± 7	172 ± 7
Weight (Kg)	58.6 ± 10	56.0 ± 8	51.4 ± 12*	60.1 ± 6	52.2 ± 7*	58.1 ± 7
Lean body mass (Kg)	49.8 ± 6	48.4 ± 6	46.2 ± 10	53.0 ± 4	47.9 ± 6*	53.0 ± 6
Haemoglobin (g dl^{-1})	14.7 ± 1.4	14.7 ± 1.2	14.9 ± 1.3m	13.2 ± 1.7	12.5 ± 1.7	12.7 ± 1.9
$\dot{V}O_2$ (l min^{-1})						
Rest	0.43 ± 0.18	0.39 ± 0.17	0.34 ± 0.19*	0.23 ± 0.10	0.21 ± 0.10	0.18 ± 0.10
Work	2.08 ± 0.71	2.08 ± 0.67	1.75 ± 0.62	1.10 ± 0.37	0.93 ± 0.25	0.80 ± 0.32*
Rest	0.61 ± 0.27	0.58 ± 0.24	0.53 ± 0.19	0.27 ± 0.11	0.23 ± 0.10	0.25 ± 0.16

* Significantly different from control values (p <0.05)

m = Eight missing values

Table 3: Laboratory tests of $\dot{V}O_2$ max predicted from $\dot{V}O_2$ at an f_H of 210-0.65 x age beats min^{-1} in absolute (Abs) terms and relative to lean body mass (LBM) and leg volume (LV). Means ± S.D.

| | Farmers | | Canal Cleaners |
	Control	Light-moderate infection	Heavy infection
n	37	147	19
Age (years)	26 ± 6	27 ± 7	27 ± 7
Height (cm)	169 ± 6	169 ± 7	172 ± 9
Weight (Kg)	57.1 ± 8	56.1 ± 8	57.1 ± 8
LBM (Kg)	49.7 ± 7	49.2 ± 6	51.7 ± 6
Haemoglobin (g dl^{-1})	15.2 ± 1.3	14.9 ± 1.2	13.3 ± 2.2*
$\dot{V}O_2$ max			
Abs (1 min^{-1})	2.71 ± 0.71	2.63 ± 0.56	2.39 ± 0.46
LBM (ml Kg^{-1} min^{-1})	54.5 ± 14.0	53.4 ± 12.5	46.2 ± 6.8*
LV (ml l^{-1} min^{-1})	247 ± 56	251 ± 55	207 ± 29*

* Significantly different from Control values ($p < 0.05$).

EFFECT OF TREATMENT ON WORKING CAPACITY

Forty-eight Gezira farmers (mean age 30 ± 12 years) with moderate to heavy *S. mansoni* infection (500-1000 eggs/g) were included in a treatment study of field work using the Oxylog method. On each of these subjects complete information was available on physiological performance, clinical status and blood biochemistry. The subjects were treated with Praziquantel (Biltricide) 40mg Kg^{-1} given in two divided doses at an interval of 4 hours. Da Silva et al (1981) reported a cure rate of 94% of patients with *S. mansoni* followed up for 6 months or more after a single oral dose of 40 mg Kg^{-1} Praziquantel. After treatment, the subjects of the present study were instructed to avoid contact with water likely to offer renewed transmission of the disease. For a period of one year, the subjects were followed up, and stools examined periodically by using a modified Kato method (Teesdale & Amin, 1976). Those who were not cured, or had been re-infected, or had a positive blood film for malaria were excluded from the study.   Subjects who were not excreting schistosome eggs after a period of one year were then re-tested by repeating the previous Oxylog tests under comparable field conditions.   A further 17 infected Gezira farmers who had participated in the original field study and not subsequently treated, were also re-tested after one year.

A similar study had been conducted 3 or 4 years earlier on Gezira farmers from the same region (Awad El Karim et al, 1981).   In that study, predicted maximal oxygen intake was measured in laboratory exercise tests in Khartoum before and after treatment with Hycanthone (Etrenol) 3 mg Kg^{-1} body weight given as a single intramuscular injection.   Hycanthone has also been found to be both a highly effective and relatively safe schistosomicide (Salih et al, 1978). Again, after treatment the subjects were screened and maintained free of further S. mansoni infection for up to one year before re-testing.   They were, in addition, given anti-malarial (chloroquine) prophylaxis during this time.   Twenty-two infected farmers (18 to 45 years age range) were given treatment and a further 19 infected farmers were investigated without treatment.   In those who remained untreated, the mean egg excretion rate after one year had increased from 257 ± 248 to 445 ± 394 eggs/g.   Treatment was given to this group finally, at the end of the study.

A paired t-test was performed to examine the changes from the first to the second tests in both treated and untreated groups (Tables 4 and 5).   A two-sample t-test was used to compare the changes between treated and untreated groups.   In both the Praziquantel study using the Oxylog and the earlier Hycanthone investigation, schistosomicidal treatment resulted in significant improvements in work capacity compared to the untreated groups.   The treated groups also showed a higher haemoglobin concentration while the untreated showed no significant change.

DISCUSSION

Many previous estimates of the effect of eliminating schistosomiasis on productive output in developing countries imply that there are major economic benefits (Khalil, 1949; Farooq, 1967; Wright, 1972).   In mainland China where the more pathogenic S. japonicum is endemic, it was claimed that the disease could cause an average loss of 40% of an adult's capacity to work (Cheng, 1971). Subsequently, these claims have not received universal support (see Adreano, 1976; Davies, 1972; Walker et al, 1972).   Our own studies in the Sudan suggest that only in the minority of the active working population, i.e. in those with high schistosome egg loads, is there a demonstrable impairment of working capacity

Table 4:   Effects of Praziquantel treatment on $\dot{V}O_2$ of Gezira farmers performing submaximal work tests in field studies. Means ± S.D. or 95% confidence limits.

	Untreated		Treated	
	1st Test	2nd Test	1st Test	2nd Test (after treatment)
n	17	17	48	48
Age (years)	33 ± 9	34 ± 9	30 ± 12	31 ± 12
Height (cm)	170 ± 7	170 ± 7	169 ± 8	169 ± 8
Weight (Kg)	61.7 ± 6	59.4 ± 6	57.4 ± 11	59.7 ± 10
Lean body mass (Kg)	51.7 ± 4	49.8 ± 4	49.4 ± 8	50.4 ± 7
Haemoglobin (g dl^{-1})	14.5 ± 1.4	14.0 ± 1.5	14.7 ± 1.3	15.0 ± 1.2*
$\dot{V}O_2$ (1 min^{-1})				
Rest	0.33 (0.20-0.72)	0.29 (0.15-0.47)	0.39 (0.11-0.84)	0.47* (0.18-0.90)
Work	1.83 (0.55-3.85)	1.73 (0.74-3.12)	2.08 (1.48-2.79)	2.36* (1.30-3.74)
Rest	0.52 (0.12-1.20)	0.47 (0.16-0.92)	0.60 (0.22-1.19)	0.57 (0.21-1.11)

*Significant difference associated with treatment ($p < 0.05$).

and physiological performance. For the most part, in field studies, infected workers appear able to work as well as the non-infected given the motivation and the same degree of skill. It suggests that there may be a threshold level of infection below which there is little or no apparent detrimental effect (Collins et al, 1976; Awad El Karim et al, 1980).

The results of our later field studies reported here again confirm this finding. In agricultural workers living in the Gezira villages of the Sudan, light or moderate egg loads of *S. mansoni* have a slight negative but not statistically significant effect on working capacity. Heavy infections with egg loads greater than 1000 eggs/g appear to diminish $\dot{V}O_2$ both at rest and during the performance of standard work tasks. Differences in working capacity can be referable to differences in stature and lean body mass. This cannot be the complete explanation, however, as the results presented in Table 3 make clear.

Table 5:   Effects of Hycanthone treatment on predicted $\dot{V}O_2$ max of Gezira
farmers performing bicycle ergometer tests. $\dot{V}O_2$ max calculated
from submaximal work loads in absolute terms and relative to lean
body mass (LBM) and leg volume (muscle plus bone) (LV). Means ±
S.D.

| | Untreated | | Treated | |
	1st Test	2nd Test	1st Test	2nd Test (after treatment)
n	19	19	22	22
Age (years)	29 ± 8	30 ± 8	25 ± 5	26 ± 5
Height (cm)	167 ± 7	166 ± 6	167 ± 8	168 ± 8
Weight (Kg)	57.5 ± 8	57.7 ± 7	52.7 ± 6	56.0 ± 8
LBM (Kg)	50.1 ± 7	50.9 ± 6	52.7 ± 6	56.0 ± 8
Haemoglobin (g dl^{-1})	14.8 ± 1.3	15.3 ± 1.1	14.1 ± 0.9	15.2 ± 0.9*
$\dot{V}O_2$max				
Abs (1 min^{-1})	2.7 ± 0.7	2.6 ± 0.6	2.4 ± 0.5	2.8 ± 0.7
LBM (ml Kg^{-1}min^{-1})	53.9 ± 12.7	51.1 ± 8.4	50.3 ± 12.9	55.8 ± 11.2
LV (ml l^{-1}min^{-1})	269 ± 46	252 ± 76	225 ± 50	245 ± 55*

*   Significant difference associated with treatment.   ($p < 0.05$)

One important factor to consider is the effect of parasitic diseases such
as schistosomiasis on haemoglobin levels.   The deleterious effects of even mild
anaemia on physical working capacity has been demonstrated both clinically and
experimentally (Viteri & Torun, 1974).   Low haemoglobin values in subjects with
high schistosome egg loads has also been reported (Nelson, 1958) and the
mechanism of haematological damage in S. mansoni has been investigated
(Mahmoud & Woodruff, 1972).   The results of work assessment and treatment
studies shown in Tables 2 and 5 amply confirm the important role played by
haemoglobin levels in all these findings.   Bengtsson and his colleagues (see this
volume) in studies on malaria in Liberia report no significant differences in work
capacity in a holoendemic group with haemoglobin levels of 14.8 ± 1.5 g dl^{-1}
compared to a mesoendemic group with haemoglobin levels of 15.3 ± 1.2 g dl^{-1}.
Such differences in haemoglobin levels may be significant ($p < 0.05$) but too small
to affect physical working capacity.   Compensatory mechanisms such as
increased circulation through energy-demanding tissues and more efficient
removal of available oxygen from haemoglobin may enable mildly anaemic
individuals to perform well at submaximal tasks.   Functional limitations may

only be revealed at near-maximal levels of exercise. Attention has also been called to the possible influence on muscle and other tissues of toxins or antigens spread systematically in chronic parasitic diseases (Bengtsson et al, this volume). There is little evidence to show how this might affect physiological function in the performance of work. Such actions, if they exist, may be small in comparison with the effects of decreased body stature and haemoglobin concentration on the cardio-respiratory parameters of work. The results of the studies reported here on a sample of over 400 agricultural workers in an endemic region of *S. mansoni* in the Sudan, show that there is little effect of light and moderate infection as quantified by egg excretion rates, but with high egg loads there is a significant reduction in power output whether this is assessed by submaximal ergometer tests and predicted $\dot{V}O_2$ max or by self-paced work in the field using the Oxylog method. Improvement of potential working capacity by schistosomicidal treatment suggests that chemotherapy may play an important role in improving physiological performance and work output.

ACKNOWLEDGMENT

This work has been supported by grants from The Edna McConnell Clark Foundation, New York.

REFERENCES

Abdel-Rahaman, T.A. (1984). Energy expenditure and schistosomiasis infection in agricultural workers of the Sudan. Ph.D. thesis, University of Khartoum.

Adreano, R.L. (1976). The recent history of parasitic disease in China: the case of schistosomiasis, some public health and economic aspects. International Journal of Health Services, **6**, 53-68.

Awad El Karim, M.A., Collins, K.J., Brotherhood, J.R., Dore, C., Weiner, J.S., Sukkar, M.Y., Omer, A.H.S. & Amin, M.A. (1980). Quantitative egg excretion and work capacity in a Gezira population infected with Schistosoma mansoni. American Journal of Tropical Medicine and Hygiene, **29**, 54-61.

Awad El Karim, M.A., Collins, K.J., Sukkar, M.Y., Omer, A.H.S., Amin, M.A. & Dore, C. (1981). An assessment of anti-schistosome treatment on physical work capacity. Journal of Tropical Medicine and Hygiene, **85**, 65-70.

Belyavin, A.J., Brown, G.A. & Harrison, M.H. (1981). The Oxylog. An evaluation. Report No. 608, Institute of Aviation Medicine, Farnborough.

Cheng, T.-H. (1971). Schistosomiasis in mainland China. A review of research and control programs since 1949. American Journal of Tropical Medicine and Hygiene, **20**, 26-53.

Collins, K.J. (1983). Energy expenditure, productivity and endemic disease, In: G. A. Harrison (ed.), Energy and Effort. SSHB Symposium Volume No.22, pp. 65-84. London: Taylor & Francis.

Collins, K.J., Brotherhood, J.R., Davies, C.T.M., Dore, C., Hackett, A.J., Imms, F.J., Musgrove, J., Weiner, J.S., Amin, M.A., Awad El Karim, M.A., Ismail, H.M., Omer, A.H.S. & Sukkar, M.Y. (1976).   Physiological performance and work capacity of Sudanese cane cutters with Schistosoma mansoni infection.   American Journal of Tropical Medicine and Hygiene, **25**, 410-421.

Da Silva, L.C., Sette, H., Christo, J., Christo, C.H., Saez-Alquezar, A., Carneiro, C.R.W., Lancet, C.M., Ohtsuki, N. & Raia, S. (1981).   Arzneim - Forsch./Drug Research, **31**(1), No. 3a.

Davies, C.T.M. (1972).   The effect of schistosomiasis, anaemia and malnutrition on the responses to exercise in African children.   Journal of Physiology, **230**, 27P.

Dounis, E., Steventon, R.E. & Wilson, R.S.E. (1980).   The use of a portable oxygen consumption meter (Oxylog) for assessing the efficiency of crutch walking.   Journal of Medical Engineering and Technology, **4**, 296-298.

Farooq, M. (1967).   Progress in bilharziasis control: Egypt.   Chronicle of the World Health Organisation, **21**, 175-184.

Forsyth, D. M. & Bradley, D.J. (1966).   The consequences of bilharziasis. Medical and public health importance in north-west Tanzania.   Bulletin of the World Health Organisation, **34**, 715-735.

Gelfand, M. (1963).   The clinical features of intestinal schistosomiasis in Rhodesia.   Central African Journal of Medicine, **9**, 319-327.

Humphrey, S.J.E. & Wolff, H.S. (1977).   The Oxylog.   Journal of Physiology, **267**, 12P.

Khalil, M. (1949).   The national campaign for the treatment and control of schistosomiasis from the scientific and economic aspects.   Journal of the Royal Egyptian Medical Association, **31**,817-856.

Mahmoud, A.A.F. & Woodruff, A.W. (1972).   Mechanisms involved in the anaemia of schistosomiasis.   Transactions of the Royal Society of Tropical Medicine and Hygiene, **66**, 75-84.

Nelson, G.S. (1958).   Schistosoma mansoni infection in the West Nile district of Uganda.   Part IV: Anaemia and S. mansoni infection.   East African Medical Journal, **35**, 581-586.

Omer, A.H.S. & El Din Ahmed, N. (1974).   Assessment of physical performance and lung function in Schistosoma mansoni infection.   East African Medical Journal, **51**, 217-222.

Salih, S.Y., Marshall, T.F. de C. & Radalowicz, A. (1979).   Morbidity in relation to the clinical forms and intensity of infection in S. mansoni infections in the Sudan.   Annals of Tropical Medicine and Parasitology, **73**, No. 5.

Shephard, R.J. (1955).   A critical evaluation of the Douglas bag technique. Journal of Physiology, **127**, 337-366.

Teesdale, C.H. & Amin, M.A. (1976).   Comparison of the Bell technique, a modified Kato thick smear technique, and digestion method for field diagnosis of Schistosoma mansoni. Journal of Helminthology, **50**, 17-20.

Viteri, F.D. & Torun, B. (1974).   Anaemia and work capacity.   Clinics in Haematology, **13**, 609-626.

Walker, A.R.P., Walker, B.F., Richardson, B.D. & Smit, P.J. (1972).   Running performance in South African Bantu children with schistosomiasis. Tropical Geographical Medicine, **24**, 347-352.

World Health Organisation (1967). Measurement of the public health importance of bilharziasis. Technical Report Series, No. 349.

Wright, W.H. (1972). A consideration of the economic impact of schisto-somiasis. Bulletin of the World Health Organisation, **47**, 559-566.

RESEARCH MODELS IN TROPICAL ECOSYSTEMS

BIOSOCIAL CONSEQUENCES OF ILLNESS AMONG SMALL SCALE FARMERS:
A RESEARCH DESIGN

R. BROOKE THOMAS[1], T. L. LEATHERMAN[1], J. W. CAREY[1] and J. D. HAAS[2]

[1]*Department of Anthropology, University of Massachusetts, Amherst, U.S.A.*
[2]*Division of Nutritional Sciences, Cornell University, Ithaca, New York, U.S.A.*

HUMAN BIOLOGICAL RESEARCH IN THE TROPICS

Widespread recognition of deteriorating conditions in the tropics has focused attention on the rapid, interdependent and frequently irreversible changes taking place throughout this region. Among these changes are a growing susceptibility of human populations to the effects of drought, deforestation, degradation of mountain resources, disease, and demographic shifts as people attempt to migrate to better conditions. Such changes have not only eroded critical relationships between human groups and their environment, but have reordered socioeconomic relationships within and between groups. As a consequence, people throughout the tropics are facing growing challenges to their self sufficiency, and in increasing numbers are trying to subsist under truly marginal conditions.

Although these conditions appear most acute in tropical Africa, they apply to large areas of Latin America and Asia as well. In comparing indicators of well-being (e.g. morbidity, mortality, undernutrition, rural out-migration, per capita food production, or foreign dependency) in the Tropical World with any other major region, the immensity of the problem is readily discernible (Buchanan, 1964; W.H.O., 1979, 1980; World Bank, 1980a,b; U.N., 1982; Pellet, 1983). In fact, the borders of the Tropical and Third Worlds are conspicuously similar. This suggests that either unique and serious impediments to meeting basic biological and social needs operate upon each of the varied tropical environments, or external political economic conditions are exerting a similar set of pressures across the tropics and on its people. Market penetration of local economies leading to the greater commercialisation of land and labour, and the ensuing rise of consumption aspirations are strongly implicated here. The selling of community food stores in drought prone areas, increased agricultural burden placed on women and children when husbands and older sons sell their labour to distant markets, the erosion of cooperative networks in agricultural communities as people leave, and the expansion of export oriented crops at the expense to small scale food production, all serve as examples.

In addressing this complex problem, an overriding concern appears to be the plight of small scale, subsistence level farmers and herders. As stated in a recent report entitled "Africa Tomorrow: Issues in Technology, Agriculture and Foreign Aid" (OTA, 1985), such producers have been consistently overlooked by both national governments and foreign assistance programs. It is not only the large number of people who fall into this category which provokes concern, but also the limited options available should they be forced to leave the land, and their disruptive influence on other aspects of the interconnected tropical system as they try to relocate (*see also* Smith et al, 1984).

Assuming that conditions affecting small scale agriculturalists will intensify beyond the "Decade of the Tropics" and well into the next century, it is appropriate to ask how non-local expertise might be used to help alleviate some of the more pervasive problems. If this is to be effective, we will need to provide specific advice concerning critical and widespread problems, and a means by which such advice can be made relevant to people living in a diversity of cultural and socio-economic settings. Our lack of success in the latter area is all too apparent, suggesting we have not fully comprehended the complexity of the cultural and environmental matrix within which recommendations are placed.

What seems appropriate, then, are approaches which are able to identify relevant problems, those sub-groups at greatest risk, and the key variables or relationships which pose limiting conditions to these highly stressed groups. Such evaluation will undoubtedly rely upon diverse methodological tools capable of determining changes in cultural, socioeconomic and biological well-being. In addition, it will need to assess the relative consequences of alternative paths of action. Given the urgency of the situation and the limited resources available to forge solutions, decisions need to be made as to areas where precise measurement is necessary, and where approximations will serve just as well. In the same sense we will need to better understand the extent to which detailed work on biosocial interactions in one population can be reliably extrapolated from those of a similar culture and social organisation or beyond.

Table 1 reviews major environmental and economic problems in the tropics, and their expected human biological consequences. Here we are particularly concerned with linkages between degree of biological impairment,

Table 1: Tropical problems and methodological recommendations

Environmental and economic problems	Human biological problems	Methodological recommendations
1. Desertification and drought	1. Nutritional inadequacy and its effects on functional impairment and recovery rates	1. More accurate assessment of biological functions directly affecting activity
2. Degradation of mountain resources	2. Health impairment and its effects on performance of essential tasks	2. Approaches which better integrate biological and social aspects of work
3. Deforestation	3. Growing work requirements and declining working potential	3. Interdisciplinary approaches by which human biologists, social scientists and local participants can better anticipate the biosocial consequences of change
4. Disease	4. Population-resource imbalances and their effects on work requirements	
5. Demographic shifts		

and the ability to function in an economically and socially productive manner. Stated differently, priority is placed upon understanding how one's biological condition (biological functional capacity) affects not only the ability to carry out essential economic tasks, such as farming and family care, but social obligations as well.   This acknowledges the critical role played by social responses in supplementing the individual and household behaviour.

With regard to the impact of these problems on human biology, better information is needed in four interrelated areas.   First, it is important to develop a clearer understanding of how different types and patterns of nutritional inadequacies influence components of biological functional capacity, such as strength, endurance and working capacity.   For instance, how long does it take for one's ability to perform strenuous work to recover to normal following chronic starvation?   Secondly, health impairment suggests a decrease in biological functional capacity resulting either from singular or multiple sources, including a combination of undernutrition and disease.   Here, questions focus on the extent to which a specific disorder, or illness in general, interferes with the performance of essential tasks.   What happens to family nutrition when the mother is chronically sick or the father breaks his leg; and is the latter as severe as a bout of pneumonia?   Also at the community level, what are the effects of an influenza epidemic coinciding with the harvest as opposed to the non-growing season?   Answers to such questions which compare the severity of a disorder in terms of its potential disruption on production have direct implications as to how limited health care options can be most effectively allocated.

On a more general level, we have already noted the growing work requirements of women as they take over tasks of their absent husbands.   A third problem area, therefore, is concerned with the consequences of overwork on women's health, and the effects this has on child care (e.g. breast feeding).   For instance, what activities are compromised in taking on this burden?   A related area of concern for human biologists is one which integrates demographic and human ecological issues with that of work.   As firewood or water for livestock become more difficult to obtain, we are interested in time and energy expenditures in order to obtain essential resources.   With this information it is possible to project the consequences of even greater scarcity or unpredictability on production and health, and to recommend which resources require priority attention.

Methodological issues which arise in asking the above questions seem not to lie so much on a greater refinement of our field techniques, but how to integrate them into a broader scope of interpretation.   We presently have available a wide range of indicators for assessing biological well-being. The challenge, then, is to select those measures which are relevant to the performance of necessary activities. For example, for individuals principally involved in low or moderate work level tasks, a low maximum aerobic capacity is of less consequence than some measure of work days lost from illness. In short, measures of biological well-being or impairment should be those which are linked in an obvious way to daily activities. This means that the human biologists must take the time to evaluate how biological functional capacity might have an effect on production, rather than assuming the universal relevance of a particular test.

Having accomplished this we are recommending a further step into the social arena. Here the linkage between individual biological functioning and social behaviour needs careful clarification. Two aspects affect the extent to which biological impairment can have a significant impact on production. First is the degree to which social responses at the household or community level effectively buffer such impairment. The second point relates to ability of the individual to reciprocate this assistance, and thus preserve it as a future resource. If either one's biological condition (e.g. chronic TB or river blindness) or over-burdened work pattern prohibits such reciprocity, the consequences of future impairment are likely to be more severe. It therefore seems appropriate that greater methodological concern be focused upon assessing the role that social behaviour plays in modifying the actual severity of biological impairment.

A final recommendation in developing a more realistic view of how human biological problems affect peoples' lives builds on the two previous methodological recommendations. As we begin to better understand relationships between biological functional capacity and production within the context of social responses, we are then in a position to anticipate some of the consequences of change which will adversely influence tropical peoples. This is clearly an interdisciplinary endeavour involving human biologists, social scientists and local participants, and one which can employ the techniques of simulation modelling. The advantage of modelling is that it offers a set of analytical procedures which build off, but supersede, the analysis of two-three variable (dependent-independent) relationships. As such, they better represent the complex interconnectivity of many variables operating within a system, and serve as a technique whereby the consequences of change as well as the efficacy of responses can be evaluated.

Applying modelling techniques to the aforementioned problems therefore offers an opportunity to anticipate some of the limiting conditions arising from

growing resource scarcity or an erosion in the cooperative network of small scale
producers. When relationships between biological impairment, working potential
and productivity are added to the model, one is in a position to better anticipate
the systemic consequences of undernutrition, illness or overwork. Furthermore,
the complex consequences of change become more readily discernible in a
biosocial sense.

Having reviewed methodological recommendations which are intended to
broaden the scope of human biological inquiries, it may be of value to provide an
example of a research design which attempts to incorporate these guidelines.
Such a design is being employed in research, supported by the U.S. National
Science Foundation (NSF BSN-8306 186), on small scale herders and farmers in
the *altiplano* region of southern Peru.   Our goals are to evaluate the
consequences of different patterns of human illness on the performance of
critical tasks, and to assess the effectiveness of social assistance in buffering
these consequences.

BACKGROUND TO ANDEAN RESEARCH

At present, Andean peasant groups are undergoing dramatic changes of
the sort already decribed. Underlying the ability of small scale farmer/herders
to remain on the land and to adjust to changing conditions is the maintenance of
adequate health.   It is for these individuals, whose livelihood is primarily
dependent upon their own physical exertion, that poor health has profound
consequences.   In this context, we are particularly interested in assessing the
consequences of an eroding cooperative network, which is expected to make the
performance of many agricultural tasks more difficult, place an increased
burden on women's work, and exacerbate the consequences of illness.   The
general relationships within such research are fairly obvious.   Illness should
reduce productivity, and assistance will help alleviate some of the consequences
of illness.   What is not apparent, and therefore of interest, is how does variation
in patterns of illness affect people's ability to get by?   Thus, we are asking what
are the relevant illness patterns, who in the population are most adversely
affected, and what are the limiting conditions beyond which these "at risk"
subgroups can no longer adjust.   The research design therefore provides a means
of narrowing a pervasive problem (illness among small scale producers) to its
most relevant components.   It thereby permits detailed inquiries on a set of
defined relationships where solutions may be more feasible or necessary.

Although many regions would have been suitable for this investigation, it
was proposed that the study be carried out in farming/herding groups residing at

4000m elevation on the highland plateau or *altiplano* of southern Peru. The selection of this region was justified for the following reasons: (1) Conditions of undernutrition (Ferroni, 1982; Picon Reatigui, 1976) and disease (Mendizabal & Cornejo, 1981; Ministerio de Salud, 1980; Dutt & Baker, 1978; Way, 1972; Buck et al, 1968; Baker & Dutt, 1972) capable of significantly disrupting productive activities are both prevalent and well documented at national, regional and local scales; (2) Extensive background data exist on biological working capacity and other biological and demographic parameters for *altiplano* populations (Baker & Little, 1976), much of it from the site of the present investigation (the District of Nuñoa); (3) Agricultural activities of peasant groups in this area are dependent on high human labour inputs—work levels of critical activities are frequently prolonged and strenuous (Thomas, 1973). Seasonal production of multiple resources puts severe constraints on household labour resources during planting and harvesting. This and the unpredictable nature of both the environment and economic conditions has led to well-developed social and economic reciprocity between households (Guillet, 1981; Orlove & Custred, 1980; Orlove, 1977; Alberti & Mayer, 1974; Escobar, 1967), forming the basis of social responses to illness.

In short, the research addresses a highly relevant problem facing small scale agriculturalists in the Tropical World. Although detailed studies have been conducted on aspects of the interrelationship between undernutrition, disease and productivity, we are very much in need of more general systemic models which can be applied to a wide range of conditions. The research scope is then necessarily broad, looking for significant interactions among critical components of the population's biological, social and economic systems rather than scrutinising singular aspects of poor health and productivity. Exploring these interrelationships in a "demonstration" population and developing a general model is but a first step. We have selected a region where appropriate conditions exist, and a study population with extensive background data to build on is available.

The research therefore examines the interaction of three critical components of small scale agricultural systems: health, agricultural work, and the social relations linking health and work. The following sections present a brief documentation on known and expected interactions between these variables. Figure 1 illustrates the scope of the investigation.

Health, illness, disease

One way in which social responses to poor health affect the illness experience, and its ultimate impact upon work, is through the translation of

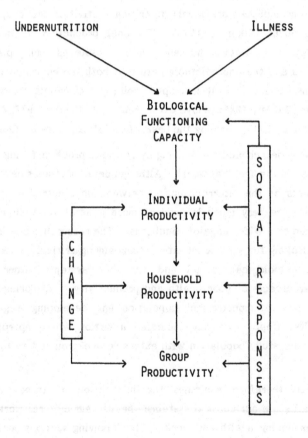

Figure 1:   Schematic diagram of interactions among factors affecting group productivity.

"insults" (Audy, 1971) into culturally identifiable illness states. In recognition of this interaction, it is important to specify what kinds of biological constraints we are examining in our studies of health, and how we define these constraints.

"Health" is viewed as a continuous process reflecting a phase of well-being and ability to function within the social, biological and environmental systems of a population, as well as to respond to future stress. Although it can be measured objectively in terms of functional impairment, the subjective interpretation of health by the population affected ultimately decides both the impact of and responses to poor health.

We are interested in "disease" as a phase in the health process "in which the ability to cope is lowered" (Armelagos et al, 1978). Disease is biomedically identifiable "abstracted out of social behaviour" by the researcher (Fabrega, 1975), and thus serves as a common denominator with other biomedical research. Our focus, however, is also on "illness" as it reflects peoples' perception of their health and for which "the culture provides an etiology, diagnosis and treatment" (Rubel, 1966). The terms "health", "disease" and "illness" share some ambiguity in meaning and are frequently used interchangeably in the health literature. The manner in which they are operationalised here is not meant as a best definition (or that they exist as distinct entities) but, rather a reflexion of the need to take into account social perceptions and practices relating to biological insult.

Health, working potential and productivity

Undernutrition and disease create a synergistic relationship making their combined effects greater than the additive effects of each (Scrimshaw et al, 1968; Gordon, 1976; Kreusch, 1975; Taylor & DeSweemer, 1973). Operating together, these factors pose the greatest health risk to small scale farming populations in "developing" countries (Winick, 1980; Latham, 1975). Thus, although we choose to focus our research on disease and illness, we do so understanding that a measure of health necessarily includes a measure of nutritional status due to its inextricable ties to the disease process.

Just as undernutrition and disease contribute to the health status of farming populations, both can and do affect biological working capacity (Barac-Nieto et al, 1978; Davies, 1974; Wheeler, 1980; Weiner, 1980). The work of Rode and Shepard (1971) is particularly interesting within the context of this study as they found that individuals having had chronic tuberculosis suffered a 12% reduction in maximal working capacity (V02 max). Biological working capacity (specifically V02 max) has in turn been linked to individual productivity in wage labourers (Basta et al, 1979; Spurr et al, 1977; Davies et al, 1973;

Hanson, 1964). The work of Collins et al (1976) in the Sudan provides one of the more complete studies linking health, working capacity and productivity. Although men heavily infected with schistosomiasis suffered a 20% reduction in working capacity, their productivity levels were equal to or greater than other members of the community. These and similar findings in the Lesser Antilles (Baldwin & Weisbrod, 1974) suggest that there are behavioral means for compensating for a physiological impairment. Thus, poor health may affect both biological working potential and individual productivity, but social compensatory responses may intercede between the linkage of work and productivity.

Social relations in illness and production

Rural populations are undergoing constant change that is increasing the degree differentiation among peasant households in their access to resources necessary to meet the tasks of farming. Key to the operation of their strategies oriented toward maintaining access to land, labour and the material resources are the formation of inter-household links of personal social networks. These links promote a diversification in the resource base and information exchanged among the various production units. They also provide access to labour needed during critical periods of the production cycle, and for tasks households cannot accomplish alone (Mayer, 1977). Since ill health can limit a household's access to its own labour and ultimately to the material resources of production, we expect interhousehold links to provide a range of options for coping in unhealthy families. Thus, illness may serve as a rallying point for group social interaction, or as an advantage point for some members of the group over others. Since the way this plays itself out in social responses used in coping with illness influences both production and future social relations, it becomes an important aspect to our research design.

STATEMENT OF PROBLEM

As previously stated, the purpose of the research has been to examine the interrelationships between health and the ability to engage in agricultural activities among small scale *altiplano* farmers and herders. Such findings should contribute to the broader goal of specifying limiting conditions under which illness can seriously disrupt the ability of rural peasants to perform essential activities. "Limiting conditions" refer to times of the year, members of households, and types of households where illness most adversely affects production, as well as to situations where social responses to illness are inadequate or not forthcoming.

Four specific objectives of the investigation are:

1. to compare morbidity and mortality patterns, and subpopulations at greatest risk in Nuñoa with other *altiplano* data sources, in order to evaluate the representativeness of the Nuñoa population;

2. to determine the degree to which ill health is associated with nutritional status, work patterns, microenvironmental risk factors, socioeconomic status and other cultural exposure conditions;

3. to assess the disruptive consequences of illness patterns on seasonal productive activities, different members of the household and household size-composition;

4. to evaluate the relative effectiveness and limitations of social responses to different patterns of illness.

RESEARCH DESIGN

Two factors have guided the design of the research. First is establishing the representativeness of the principal variable under investigation within the population and region, e.g. how important is illness in Nuñoa and to what extent does it reflect patterns of other *altiplano* areas? Secondly, is gathering sufficiently detailed information on relationships between patterns of illness, and consequences and responses to illness, in order to understand this as an interactive process susceptible to change? The former relies upon aggregate data sources which already exist at the department, provincial and district level, and which will be augmented locally through extensive surveys. This information, in turn, provides a foundation for intensive inquiries into the more disruptive types and patterns of ill health. Findings from these two modes of inquiry can then be integrated into our present knowledge of the Nuñoan and other *altiplano* agricultural systems, and analysed with regard to future changes capable of impacting disease or the performance of agricultural activities.

Each data collection phase builds upon information provided in the previous phase. Data collection commences with large sample scanning of household health problems, and background data, using a preliminary health survey. Progressively more detailed inquiries follow, employing a series of nested subsamples derived from the preliminary sample. Nested sampling, therefore, allows us to accumulate an extensive information base on specific households, as well as to assess the representativeness of these sub-samples relative to the population. Data collection phases in the research design follow.

Aggregate data analysis.   Using government data sources, morbidity-mortality patterns and subpopulations at highest risk at the department, province and district levels are identified.

Preliminary health survey.   Random or representative samples are drawn from three major residence types in the District duing the planting/wool shearing season in order to provide preliminary data on variability in health history, recent symptoms, and associated variables.

Seasonal surveys.   Using randomly selected subsamples from the preliminary survey, two additional surveys are conducted corresponding to the harvest and post-harvest seasons. Questionnaires pertaining to symptomatology, morbidity events and household productivity over a three-month interval are administered.

Supplementary surveys.   A series of four additional surveys are used to amplify information gained in the preceding surveys.   They address: (1) micro-environmental factors capable of influencing illness;   (2) identification of a household's social support network, and assessment of household socioeconomic status;   and (3) hypothetical responses to different types and patterns of illness. One of these supplemental surveys is administered along with each of the two seasonal surveys.

Time allocation case studies.   Households of ill individuals are observed at intervals throughout their illness and into recovery in order to collect detailed contextual and time expenditure data on activity pattern.   Samples, when possible, are drawn from the seasonal survey with the intent of comparing four categories of illness: acute moderate and severe, and chronic moderate and severe.   Since this phase of data collection is designed to amplify results, survey findings specifying limiting conditions guide sampling considerations.

Working potential tests.   A series of standardised work physiology tests are given to moderately ill and recently recovered individuals in order to assess the effects of illness in reducing biological work potential.   Tests consist of moderate and strenuous agricultural activities, as well as those of submaximal and (pending a physician's approval) maximal work capacity.   Here we are concerned with assessing long-term effects of respiratory disease.   Samples consist of young-middle aged men engaged in agricultural activities. Comparisons are made between three groups of moderately ill, recently recovered, and healthy individuals without a history of respiratory disease.

Population health history.   Major health-related events, such as epidemics, which have taken place in the Nuñoa District are described in detail by the local health officer and other key informants. Attention is given to community consequences of these events and to extreme cases where households have had to disband or migrate. We do not anticipate that this information will be elicited in sufficient detail from other data collection phases.

SAMPLING CONSIDERATIONS

Since ultimately inference of the impact of health variation is made at the household and community level, these serve as primary and secondary sampling units for our survey. We have chosen three significantly different residence units based on subsistence type and social organisation. These factors influence living conditions, and presumably health and health care systems. The residence units are: the largest town in the District (Nuñoa); an alpaca herding cooperative (Chillihua); and a more traditional community of primarily farming families (Sincata).

The town of Nuñoa is the subject of a 25 percent random sampling. It can be divided into several zones according to the municipal services available and general socioeconomic status. Each household within the town is given a number which serves to identify it if chosen from a random number table. This ensures a representative distribution of households across socioeconomic levels. Households that refuse to participate are tallied after the first survey, and an equal number of replacement households are chosen randomly from the same respective socioeconomic zones of the town from which the nonresponders originated. A 100 percent sample of the cooperative and a 50 percent representative sample of the traditional farming community or *ayllu* is selected, providing a minimum of 25 families each.

Seasonal subsamples consist of half the households previously surveyed, although a minimum of 15 families is deemed necessary from any one residence unit for application of statistical analyses. These subsamples are representative of the larger initial health survey sample in that we attempt to select households at random from each residence unit. Nevertheless, a strict random sample will probably yield insufficient numbers of households at greatest risk to illness. Therefore, a second additional sample is chosen from the pool of families excluded by the random sampling procedure, and representing the five families of each residence unit in poorest health. This determination is made from a review of the preliminary health survey. The additional families are noted as a separate sample so that generalisations from such surveys to the larger population are based only on the randomly selected households. Sample size by residence type appears in Table 2.

Table 2: Structure of surveys

Phase I (Year 1)	Phase II (Year 2)
A. Sampling considerations: Samples drawn randomly from three different residence types in order that findings be representative of the population	Sampling focus on individuals and households directly affected by illness

Sample size:

Phase I — Residence type and number of households:

First survey		Second and third follow-up surveys	
Town (traditional)	90	Town (traditional)	60
Ayllu (agrarian)	25	Ayllu (agrarian)	20
Coop	25	Coop	20
TOTALS	140		100

Phase II —

1. Time allocation case studies (households)

Illness pattern	Men	Women
Acute moderate	6	6
Acute severe	6	6
Chronic moderate	6	6
Chronic severe	6	6
Sickest homes	5	5
Healthiest homes	5	5
TOTALS	34	34

2. Working potential tests (individuals)

Illness pattern	Men
Chronic moderate resp. disease	30
Recently recovered	30
Healthy w/o resp. disease history	30
	90

B. Method and type of data collected:

Phase I

1. First Survey (Planting season)

Illness symptomatology survey
Anthropometrics
Household demographic survey
Illness history survey
Structured observations of house type, dress, etc.
Health habits and food consumption survey

2. Second Survey (Harvest season)

Illness symptomatology survey
Anthropometrics
Household production activities survey
Observations and measures of micro-environmental risk factors
Social support network questionnaire

3. Third Survey (Post-Harvest season)

Illness symptomatology survey
Anthropometrics
Household harvest activities survey
Socioeconomic status determination using observation and key informants
Hypothetical responses to illness patterns
Cultural perceptions of illness

Phase II

1. Time allocation case studies

Time-motion analysis on activities of households with ill adults conducted throughout the illness period and into recovery.
Anthropometrics

2. Working potential tests

Oxygen consumption and heart rate measured during standardised agricultural, and submaximal working capacity tests.
Strength tests
Anthropometrics

3. Population health history

Key informant recall of major health events and extreme cases in Nuñoa's past

In recommending a random sampling procedure, we are aware how this can be distorted under field conditions. When a substantial portion of the selected sample refuses to participate, substitution tends to introduce biases favouring over-representation of lower socioeconomic households. Whereas, the rapport established in the area diminishes this effect, we have attempted to substitute with similar status households. Finally, should such a bias occur, it generally results in an over-reporting of illness in the population, assuming wealth and health are associated (Bolton & Sue, 1981). Since data collected from the preliminary survey permits detection and correction for this, the direction of the potential bias is not disadvantageous to the design of the research.

METHODS

Specific methods pertaining to data collection phases appear below.

Preliminary health survey

The preliminary survey consists of the following: household demographic information, individual illness history and symptomatology, anthropometrics and structured observations on living and socioeconomic conditions.

1. A *household demographic questionnaire* follows those used in previous studies in Nuñoa. Information sought includes family size, sex-age composition, fertility-mortality history, birthplace, residence changes, occupation, education, marital status and migration history.

2. A *health history questionnaire* relies upon recall techniques to determine the manner and extent to which illness has affected households members throughout their life cycle. Questions will be open-ended and focus upon disorders known to be prevalent in Nunoa (Spector, 1971; Way, 1972) as well as regional perceptions of illness (Frisancho, 1972; Stein, 1981). Household members are asked to organise major illnesses or accidents by stages of life: early childhood (0-5 years), late childhood (6-12 years), adolescence (13-20), young-middle aged adult (20-45), older adult (46-60). They are asked to describe illness events within the last three months, one year and five years. For consistency, these are administered to mothers who are expected to have a more accurate recall of their health history.

Whereas recent recall information on illness is undoubtedly more detailed and accurate, life stage recall can elicit a record of the most dramatic events and their consequences. This in turn prepares the informant for more specific questions on symptomatology.

3.   The *health symptomatology questionnaire* is designed to assess existing rather than retrospective health status.   As Dutt and Baker (1978) have noted, this technique is selected because past research has repeatedly indicated that such questionnaires yield more information on symptoms and a better presumptive diagnosis than that obtained from most clinical histories taken by physicians (Brodman et al, 1951; Scotch & Geiger, 1964; Abramson et al, 1965). Such questions provide fairly accurate health profiles at the population level, and allow the isolation of specific body systems under stress through identification of constellations of symptoms.   Whereas the bulk of the questions require yes/no responses, we encourage the recording of amplifying comments. Questions focus on symptoms relating to one or more descriptive domains of disease and illness, overall health, chronic problems, acute problems, musculo-skeletal problems, parasitic load, respiratory disorders, circulatory and heart problems, gastro-intestinal disorders, and psycho-social stress. Follow-up questions eliciting perception of severity and duration of a problem are interspersed among symptom domain responses.

4.   *Anthropometric assessment* of nutritional status is made following Weiner and Lourie (1969) and includes the following measures: stature, weight, sitting height, upper arm circumference, chest circumference, biacromial diameter, bi-iliocristal diameter, head circumference (for children), and triceps and subscapular skinfolds.

5.   *Structured observations* on visible characteristics of house type and condition, water source, dress and other material possessions are recorded.   These serve as gross indicators of socioeconomic status, and are particularly useful for internal comparisons within a community.

Seasonal surveys

Information collected in the aggregate data analysis and preliminary survey phases permit identification of the variability in health problems within the Nuñoa population, and the degree to which patterns of disease and illness reflect regional norms.   Such a contribution is comparable to the goals of past health survey studies already cited in the region.   The thrust of the study, however, is to go beyond this level of analysis, and to examine in detail the nature of the pattern and its consequences on productive activities. In order to accomplish this, it is necessary for seasonal health and productivity surveys to be utilised.

Two such surveys are carried out on subsamples of the preliminary health survey sample. The preliminary and seasonal surveys are scheduled so to correspond to important annual production activities: planting and the early wool shearing season; the pre-harvest and harvest season when nutritional stores are lowest and weather conditions harshest; and the post-harvest season when nutrition stores are ample and much non-agricultural work (e.g. migration for wage labour) takes place.

Seasonal surveys are administered to household members and consist of the following parts: questions on morbidity recall, symptomatology, and household production as well as anthropometric measures.

1. A *morbidity recall questionnaire* is open-ended and pertains to the illness events during the three-month interval since the last survey. Respondents are asked to evaluate severity and duration of illness, how these affected performance of activities or production, who assisted them when their performance was impaired, what they did to cure the illness, and possible causes of the illness.

2. *Symptomatology questionnaire* follows the same format described in the preliminary survey.

3. A *household production questionnaire* addressing activities during the three-month interval is administered to household heads. This question-naire not only provides a context within which to discuss illness and productivity, but yields valuable information on environmental exposure pattern, critical activities performed by family members, and the working conditions under which tasks are or cannot be carried out. Questions pertain to crop and animal production, earnings from wage labour, assistance given without remuneration, household and herd maintenance, the production and preparation of textiles, fuel acquisition, and food processing and preparation. Just as illness is expected to be most disruptive during the busiest times in the year, it would have a similar effect when it impairs a person's ability to perform a critical task which no one else in the household could do. Questions therefore will attempt to uncover necessary levels of production, what has actually been produced, and what are the limiting conditions in the production process.

4. *Anthropometrics*, described earlier, are gathered on household members. These measures are used to document seasonal changes in nutritional status and the results of illness on growth and adult morphology.

Supplementary surveys

A series of four additional surveys are used to amplify information gained from the preliminary and seasonal surveys.   These address the following areas: microenvironmental risk factors capable of influencing illness, household social support network, identification of the household socioeconomic status, and hypothetical response questions to different types of patterns of illness.

One or more of these surveys is administered along with each of the two seasonal surveys.   The above order takes advantage of an accumulating data base from which to ask increasingly more specific questions, and puts off socially sensitive inquiries until the informants have become better acquainted with the investigators.   For this reason, it is desirable that the same research team be present for each survey.

1.   Although the investigation focuses upon consequences rather than causes of ill health, two *microenvironmental risk factors* appear particularly relevant to public health issues in the study area.   These are smoke exposure and water quality.   Recent reports have noted possible deleterious health effects of continuous exposure to dung smoke (Davidson et al, 1981).   This is coupled with the high frequency of respiratory disease among *altiplano* populations, and their reliance on animal dung as a fuel (Winterhalder et al, 1974).   A preliminary analysis of dung smoke indicates a high concentration of polycyclic aromatic hydrocarbons, suggesting a serious health threat.   A series of questions are administered to determine exposure time and type, quantity, and condition of fuel burned.

Contaminated water sources significantly affect health status (Chanlett, 1979).   This is particularly relevant on the *altiplano* where water quality changes dramatically with the wet and dry season, and potential human and animal contamination is expected to be severe.   Fecal contamination and specifically the presence of fecal coliform bacteria are used as an indicator of water quality, and measured using the Millipore "dipstick" procedure.

2.   *Social support network questions* probe outside assistance provided to household members.   Informants are asked to identify individuals who would help out in a number of contexts.   From such information a list of the functioning support network for a household is identified, and the frequency and context of its use specified.   This in turn provides necessary background information for examining social responses to

different patterns of illness.

3.   Whereas observational evaluation of *household socioeconomic status* is provided in the preliminary survey, we intend to substantially augment these data by two techniques.   First is a check list of material possessions reflecting status, such as metal pots, radios, bicycles, sewing machines, and size of herds.   The second technique, successfully employed by Bolton and Sue (1981), depends upon key informants to provide a relative wealth ranking of known households.

4.   *Responses to illness* fall into two categories: therapeutic responses by the "patient" and others, and responses to compensate for the patient's impaired ability to contribute to productive activities.   Whereas the purpose of the research is not to examine curing practices, time and resources spent in the caring and curing process interferes with productive activities.   The focus of this final supplemental survey is to elicit information on the range of strategies employed at the individual, household, and extrahousehold level to compensate for the effects of illness.   Here we want to understand in some detail under what circumstances are different responses utilised, and what are the perceived costs and benefits in employing them.

Questioning begins by amplifying information on experienced illnesses taken from the illness history questionnaire or seasonal surveys. Specifically we will want to know responses to different patterns of illness as they affect key participants, critical activities and households with varying composition.   To elicit this information questions will be posed in a hypothetical "what if?" manner.

Time allocation case studies

During the second year detailed, time allocation case studies build on survey recall information.   Here, focus is upon contextual and quantitative time allocation data pertaining to activity pattern of households having ill members.

Samples, when possible, are drawn from seasonal samples with the intent of comparing consequences of and responses to different patterns of illness. Patterns considered to be both relevant and amenable to investigation are those differing in extent and degree of disruption.   Thus, a minimum categorisation of illness follows four patterns: acute, moderate and severe;   and chronic, moderate and severe.   "Acute" illness is arbitrarily defined as that lasting up to one month, and "chronic" beyond that point.   "Moderate" refers to the inability to do heavy work, such as load carrying or field preparation (*see* Thomas, 1973, for workload definitions of agricultural activities), and impaired performance of sustained, moderate workloads.   "Severe" denotes that activity is restricted to sedentary actions.

Longitudinal observations are made through the period of illness and into recovery.   Our sampling goal is to follow six male and six female household heads in each of the four illness pattern categories.   Comparable information are also gathered on the five households of poorest health, studied the year before in the seasonal survey, and five of the healthiest households from the same survey. Whereas the former sample provides second year follow-up data on long-term effects of illness, the latter serves as a control (*see* Table 2).

Since the pattern of illness will not be apparent as we begin to study the household of a recently ill individual, sampling selection is guided by situations providing a clearer definition of limiting conditions.   We anticipate that illness affects productive activities most when it strikes young-middle aged adults having a families with a low producer-consumer ratio (thus our selection of household heads for study).   Critical activities should be those falling at the busiest times of the year (planting and harvest) and having the most strenuous, prolonged work demands.

Upon identifying and obtaining permission to work with an ill individual, two techniques of timed observations are conducted.   The first, following standard techniques utilised by Thomas (1973), requires continuous observation throughout daylight hours.   It provides detailed data on sequences of activities, and the interactive context they occur in.   Although time-consuming, this approach is necessary in order to initially understand a particular family's daily round.   The second technique entails recording activities observed during a randomly scheduled 30-minute visit (Johnson, 1975).   The advantage of the random visit approach is that a large number of households can be surveyed in a single day, and the participation of members in broad activity categories can be compared.

Households contacted initially are observed using the daily round technique for two days during the first week.   Thereafter we utilise the random visit approach once a week for the first month, and following the first month once a month until recovery.   With the exception of the first observational period, a set of questions designed to recall activities and responses taking place in the interim between visits are administered.   By asking the family what it would be doing differently if one of its members were not ill, the family serves as its own and most appropriate control.

As we get to know the household better, a series of questions paralleling and amplifying those given in the seasonal and supplemental surveys are administered. Timing depends on the convenience of the household, but takes place so as not to disrupt timed observations. In addition, anthropometrics are conducted on household members once a month for the remainder of the year.

Working potential tests

Physiological tests of working potential are designed to assess an individual's biological ability to perform at a range of work levels. Such information serves as an important supplement to perceptual concepts of illness collected through recall data and activity observation. Of particular interest is the extent to which perceptions and/or behavioral patterns associated with illness reflect a decrease in physiological working capacity.

Do people, for instance, over- or under-compensate behaviorally when confronted with diminished working potential? Reports from Nuñoa (Barreda, personal communication) indicate that moderate sickness is generally ignored during the busiest times of the year. Thus, despite suspected decreases in working potential (strength, endurance) it may be unlikely that an observer would notice a significantly different pattern of work because of variability in job, location and work group composition.

In order to overcome this difficulty, a series of standardised working potential tests are conducted on chronic, moderately ill individuals as well as those recently recovered. The first set of tests simulates critical agricultural activities. A second set consists of better controlled submaximal and, when possible, maximal exercise tests.

Of principal concern is the testing of recently recovered individuals and the time it takes for working potential to return to normal values. This should be particularly important for those having had chronic respiratory disease where damage to lung tissue can result in persistant interference with oxygen transport, and hence working capacity (Rode & Shepard, 1971), especially at high altitude (Velasquez, 1973). Because respiratory disease constitutes the most important single health problem among *altiplano* peoples, sample recruitment is designed to test associations between indicators of working potential and type, severity and duration of respiratory disease.

Samples are drawn from the time allocation study or seasonal survey, and consist of young-middle aged men normally engaged in agricultural work. Women will not be studied because of their reluctance in the past to participate in such tests. Two samples of 30 men each, one with chronic moderate respiratory disease, the other having recently (within the last three years) recovered, will be compared to 30 healthy men without a history of said disease (*see* Table 2).

Ill volunteers will be screened by a physician.   Those accepted will be studied doing moderate level, agricultural activities and submaximal tests only.

1.   In order to study the role of respiratory disease in limiting worker productivity under natural work conditions, a series of agricultural activities are examined.   This is accomplished using a protocol developed by Thomas (1973) for the field testing of the following standardised activities: long-distance walking, load carrying (adjusted for body weight) and potato field preparation.   For all field tests subjects are expected to maintain the same rate over a specified course.   Heart rate and expired air samples are taken preceding the test and at intervals throughout, including recovery.   Expired air samples are collected using a K-M respirometer and a Collins J-2 valve.   Oxygen consumption is determined by the Weir (1949) method using a low resistance dry gas meter and a paramagnetic oxygen analyser (Beckman C-2).

2.   Submaximal and maximal ergometric tests are conducted on a bicycle ergometer.   The testing protocol follows that previously employed by Buskirk (1976) and others in studies of Nuñoa men during the 1960's.   Two submaximal tests will be performed.   The first consists of riding the bicycle for 5 minutes at 3 kpm and 60 rpm (1080 kpm/min).   The second, for 30 minutes at 2.5 kpm and 60 rpm (900 kpm).   Heart rate and expired air samples are taken during resting (RMR) preceding exercise, and at intervals during testing and into recovery.

The maximal work test is designed to exhaust the subject in 10 minutes or less.   After a 5-minute warm-up ride at 3 kp (1080) and a 1-2-minute rest, the initial rate is retained for the first two minutes.   Thereafter, the load is increased 0.5 kp (180 kpm) per minute until the subject can no longer maintain the pedaling frequency of 60 rpm.   The highest oxygen consumption measured during the ride is recorded as maximum oxygen uptake.

Submaximal and maximal expired air samples are collected using a Collins J-3 valve connected to a series of Douglas Bags.   Oxygen consumption determinations are made with a Scholander micro-gas analyser.

Preceding testing, subjects are asked personal questions pertaining to age, occupation and disease history.   Anthropometrics and strength testing of hands, arms, and legs are then conducted using a grip strength dynometer and cable tensiometer.

POPULATION HEALTH SURVEY

The health history of the population is intended to document major health-related events, such as epidemics, which have taken place in Nunoa over the past several decades.

A local health worker is asked to amplify factors associated with different illnesses, seasonal variations, consequences to different types of households, and what people do about it. Here we probe for case studies of households who have had to split up or migrate, and thus may not be well represented in our samples. This account is verified by other key informants.

SIGNIFICANCE OF THE RESEARCH

It is our opinion that the strength of this research design lies in its biosocial scope, and in the array of both objective and subjective interpretations of health status which it attempts to integrate. As was stated in the beginning of the paper, we are extremely concerned about the deleterious consequences of social and economic change in the tropics whereby segments of populations, and sometimes whole communities, are faced with declining health, undernutrition and growing workloads. By proposing exploratory research which provides a comprehensive overview of the interaction between illness, nutritional status, work potential, production and social relations, we are provided with insights into some rather complex processes. At very least such an analysis alerts us to the degree of variability within what appears, too often, to be a fairly homogeneous peasant population. In addition, it is capable of flagging a set of limiting conditions beyond which serious degradation of critical relationships at the household or group level would be expected.

By gaining an understanding of the subgroups and relationships at greatest risk, one is in a better position to assess the consequences of change impacting communities of small scale farmers and herders. Here, simulation modelling offers some assistance as an analytical technique. While not predictive, modelling does provide insights into alternative scenarios which could occur. For example, the consequences of different patterns of illness on production might be evaluated. An influenza epidemic occurring during the planting season is expected to have very different consequences than one in the inactive post-harvest season. Likewise, the convergence of unlikely but interrelated events such as drought, epidemic disease and rising food costs is subject to analysis.

Finally, it would be possible to simulate high rates of out-migration of young men seeking wage labour opportunities, and to note the effects of a reduced cooperative network on filling in for the tasks of ill individuals. Here a constant frequency of illness is expected to have more serious repercussions with the erosion of social support networks.

In conclusion, then, we anticipate the research to identify the following: (1) environmental, nutritional and socioeconomic factors associated with disease, (2) the extent to which disease affects work and production among small scale farmer/herders; (3) the most disruptive types and patterns of disease, and subgroups most affected; (4) the effectiveness of social responses in coping with varying patterns of ill-health; (5) changes capable of increasing the frequency and consequences of disease.

These results, in turn, are expected to make the following theoretical and applied contributions: (1) to provide a basis for modelling effects of varied disease patterns on small human populations; (2) to demonstrate seriousness of health problems under seasonal and relatively unpredictable conditions; (3) to provide a basis for prioritising health problems in terms of their disruptive effects on production; (4) to produce a broad based model whereby consequences of anticipated change can be simulated; (5) to supplement knowledge of biocultural adaptation to high altitude environments.

The research design has therefore attempted to link quantitative and qualitative data sources. It begins by establishing the representativeness of the population with others on the *altiplano* through the analyses of aggregate data sources. It then looks within a population, attempting to determine the representativeness of illness by season, residential units and socioeconomic position. As the study proceeds, more detailed measures of variation in patterns of illness and work performance become the focus. The end of the progressively narrowing scope of inquiry centers upon interviewing the few individuals knowledgeable in the community's long-term health history. They are in a position of providing us with information which aggregate data sources do not show, which individuals responding to questionnaires do not recall, and which more precise measures of time allocation and working performance are too limited to yield. In summary, the diverse data sources are designed to complement one another, and in doing so provide a more accurate overview.

REFERENCES

Abramson, J.H., Terespolsky, L., Brook, J.G. & Kark, S.L. (1965). Cornell medical index as a health measure in epidemiological studies. British Journal of Preventive and Social Medicine, **19**, 103.

Alberti, G. & Mayer, E. (1974). Reciprocidad e Intercambio en los Andes Peruanos. Instituto de Estudios Peruanos, Lima, Peru.

Armelagos, G.J., Goodman, A.H. & Jacobs, K. (1978). The ecological perspective in disease, In: M. Logan & E. Hunt (eds.), Health and Human Condition. North Scituate, Massachusetts: Duxbury Press.

Audy, J.R. (1971). Measurement and diagnosis of health, In: P. Shepard & D. McKinley (eds.), Environ/Mental. New York: Houghton Mifflin.

Baker, P.T. & Dutt, J.S. (1972). Demographic variables as measures of biological adaptation: a case study of high altitude human populations, In: G. A. Harrison & A. J. Boyce (eds.), The Structure of Human Populations. Oxford: Clarendon Press.

Baker, P.T. & Little, M.A. (1976). Man in the Andes: A Multidisciplinary Study of High-Altitude Quechua. Stroudsburg, Pennsylvania: Dowden, Hutchinson & Ross.

Baldwin, R.E. & Weisbrod, B.A. (1974). Disease and labor productivity. Economic Development and Cultural Change, **22**(3), 414–435.

Barac-Nieto, M., Spurr, G.B., Maksud, M.G. & Lotero, H. (1978). Aerobic work capacity in chronically undernourished adult males. Journal of Applied Physiology, **44**(2), 209–215.

Barreda, V. (Personal communication).

Basta, S.S., Soekirman, D.S., Karyad, D. & Scrimshaw, N. (1979). Iron deficiency anemia and the production of adult males in Indonesia. American Journal of Clinical Nutrition, **32**, 916–925.

Bolton, R. & Sue, M. (1981). Health and wealth in a peasant community, In: J. W. Bastien & J. M. Donahue (eds.), Health in the Andes. AAA Publication No. 12, American Anthropology Association, Washington, D.C.

Brodman, K., Erdmann, A.J., Lorge, I. & Wolff, H.G. (1951). C.M.I. Health Questionnaire - II, as a diagnostic instrument. Journal of the American Medical Association, **145**, 152.

Buchanan, K. (1964). Profiles of the Third World. Pacific Viewpoint, **5**(2), 97–126.

Buck, A.A., Sasaki, T.T., Anderson, R.I. (1968). Health and disease in four Peruvian villages: contrasts in epidemiology. Johns Hopkins Press, Baltimore.

Buskirk, E.R. (1976). Work performance of newcomers to the Peruvian highlands, In: P. T. Baker & M. A. Little (eds.), Man in the Andes. Stroudsburg, Pennsylvania: Dowden, Hutchinson & Ross.

Chanlett, E.T. (1979). Environmental Protection (2nd ed.). New York: McGraw-Hill.

Collins, K.J., Brotherhood, R.J., Davies, C.T.M., Dore, C., Hackett, A.J., Imms, F.J., Musgrove, J., Weiner, J.S., Amin, M.A., El Karim, M., Ismail, H.M., Omer, A.H.S. & Sukkar, M.Y. (1976). Physiological performance and work capacity of Sudanese cane cutters with Schistoma mansoni infection. American Journal of Tropical Medicine and Hygiene, **25**, 410–421.

Davidson, C.I., Thomas, C.G. & Nasia, M.A. (1981).   Airborne lead and other elements derived from local fires in the Himalayas. Science, **214**, 1344-1346.

Davies, C.T.M. (1974). The physiological effects of iron deficiency anaemia and malnutrition in exercise performance in East African school children. Acta Paediatricia Belgica, **28** (Supplement), 253-256.

Davies, C.T.M., Chukwenmeka, A.C. & VanHaaren, J.P.M. (1973).   Iron deficiency anaemia: its effect on maximum aerobic power and responses to exercise in African males aged 17-40 years. Clinical Science, **44**, 555-562.

Dutt, J.S. & Baker, P.T. (1978). Environment, migration and health in southern Peru. Social Science and Medicine, **12**, 29-38.

Escobar, M.G. (1967). Organizacion Social y Cultural del Sur de Peru. Serie Antropologia Social 7. Inst. Indigenista, Interamericano, Mexico, D.F.

Fabrega, H. (1975). The need for an ethnomedical science. Science, **189**, 969-975.

Ferroni, M.A. (1982).   Food habits and the appartent nature and extent of dietary nutritional deficiencies in the Peruvian Andes. Archives Latino-americanos de Nutricion. **32**(4), 850-866.

Frisancho, D. (1972).   Creencias y Supersticiones Relacionades con las Efermedades del Altiplano Puneo. Los Andes: Puno, Peru.

Gordon, J.E. (1976). Synergism of malnutrition and infection, In: G. H. Beaton & J. M. Bengos (eds.),   Nutrition in Previtnitive Medicine,   WHO, Greece.

Guillet, D. (1981).   Agrarian ecology and peasant production in the central Andes. Mountain Research and Development, **1**(1), 19-28.

Hanson, J.E. (1964). The relationship between individual characteristics of the workers and output of work in logging operations.   Proceedings of the Second International Congress on Economics, Dortmund, 1964, pp. 113-120. London: Taylor & Francis.

Johnson, A. (1975). Time allocation in a Machiguenga community. Ethnology, **14**, 301-310.

Kreusch, G.T. (1975).   Malnutrition and infection: deadly allies.   Natural History, **84**(9), 27-34.

Latham, M.C. (1975). Nutrition and infection in national development. Science, **188**, 561.

Mayer, E. (1977). Beyond the nuclear family, In: R. Bolton & E. Mayer (eds.), Andean Kinship and Marriage. AAA Pub., No. 7, pp. 1-27. Washington: American Association of Anthropologists.

McRae, S.D. (1979). Human Adaptation and Planned Change in the High Andes of Southern Peru: A Simulation Approach. Ph.D. dissertation, Department of Natural Resources, Cornell University.

Mendizabal, L., G. & C. Cornejo Rosello V. (1981). Extension of health service coverage in Puno, Peru. Bulletin of the Pan American Health Organisation, **15**(2), 121-130.

Ministerio de Salud (1980). Enfermedades Transmisibles Peru, 1979. Oficina Sectorial de Estadistca e Informatica. Lima, Peru.

Orlove, B.S. (1977).   Inequality among peasants: the forms and uses of reciprocal exchange in Andean Peru. In: R. Halperin & J. Dow (eds.),

Peasant Livelihood Studies in Economic Anthropology and Cultural Ecology. New York: St. Martin's Press.

Orlove, B.S. & Custred, G. (1980). The alternative model of agrarian society in the Andes: households, networks, and corporate groups, In: B. S. Orlove & G. Custred (eds), Land and Power in Latin America. New York: Holmes & Meier.

O.T.A. (1985). Africa Tomorrow: Issues in Technology, Agriculture and Foreign Aid. Washington, D.C.: Office of Technology Assessment.

Pellet, P. (1983). Commentary: changing concepts on world malnutrition. Ecology of Food and Nutrition, **13**, 115-125.

Picon-Reatigui, E. (1976). Nutrition, In: P. T. Baker & M. Little (eds.), Man in the Andes. Stroudsburg, Pennsylvania: Dowden, Hutchinson & Ross.

Rode, A. & Shepard, R. J. (1971). Respiratory fitness of an arctic community. Journal of Applied Physiology, **31**(4), 519-526.

Rubel, A. (1966). The role of social science research in recent health programs in Latin America. LAAR, **2**(1), 37-56.

Scotch, N. & Geiger, H. (1964). An index of symptoms and disease in Zulu culture. Human Organization, **22**, 304.

Scrimshaw, N., Taylor, C.E. & Gordon, J.E. (1968). Interactions of Nutrition and Infection. WHO Series No.57, Geneva.

Smith, J., Wallerstein, E. & Evers, H.-D. (1984). Households and the World-Economy. Beverly Hills, California: Sage Publications.

Spector, R.M. (1971). Mortality Characteristics of a High Altitude Peruvian Population. Master's thesis in Anthropology, The Pennsylvania State University.

Spurr, G.B., Barac-Nieto, M. & Maksud, M.G. (1977). Productivity and maximal oxygen consumption in sugar cane cutters. American Journal of Clinical Nutrition. **30**, 316-321.

Stein, W.W. (1981). The folk illness: entity or nonentity? An essay on Vicos disease ideology. In: J. W. Bastein & J. M. Donahue (eds.), Health in the Andes. AAA Publication No. 12. American Association of Anthropologists, Washington, D.C..

Taylor, C.E. & DeSweemer, C. (1973). Nutrition and infection, In: M. Recheigl (ed.), Food, Nutrition and Health. New York: Karger.

Thomas, R.B. (1973). Human Adaptation to a High Andean Energy Flow System. Occasional Papers in Anthropology, No. 7. Pennsylvania State University, University Park, Pennsylvania.

United Nations (1982). Report on the World Social Situation. Department of International Economic and Social Affairs. United Nations, New York.

Velasquez, T. (1973). La Silicosis en el Peru. Instituto de Salud Occupacional.

Way, A.B. (1972). Health Exercise Capacity and Effective Fertility Aspects of Migration to Sea Level by High Altitude Peruvian Quechua Indians. Ph.D. Dissertation, University of Wisconsin, Madison, Wisconsin.

Weiner, J.S. (1980). Work and well-being in savanna environments: physiological considerations, In: D. R. Harris (ed.), Human Ecology in Savanna Environments. London: Academic Press.

Weiner, J.S. & Lourie, J.A. (eds.) (1969). Human Biology: a Guide to Field Methods. IBP Handbook, No. 9. Oxford: Blackwell.

Weir, J.B. de V. (1949).   New methods for calculating metabolic rate with special reference to protein metabolism. Journal of Physiology, **109**, 1.

Wheeler, E.F. (1980).   Nutritional status of savanna peoples, In: D. R. Harris (ed.),   Human Ecology in Savanna Environments.   London: Academic Press.

W.H.O. (1979).   Formulating Strategies for Health for All by the Year 2000. Geneva: World Health Organization.

W.H.O. (1980).   Sixth Report on the World Health Situation.   Part One - Global Analysis. Geneva: World Health Organization.

World Bank (1980).   Health Sector Policy Paper. Washington: The World Bank.

World Bank (1980).   Meeting Basic Needs: an Overview.   Washington: The World Bank.

Winick, M. (1980).   Nutrition in Health and Disease. New York: Wiley.

Winterhalder, B., Larsen, R. & Thomas, R.B. (1974).   Dung as an essential resource in a highland Peruvian community. Human Ecology, **2**, 89-104.

THE TEA PLANTATION AS A RESEARCH ECOSYSTEM

D. J. BRADLEY[1], L. RAHMATHULLAH[2] and R. NARAYAN[3]

[1] *Ross Institute and Department of Tropical Hygiene,*
London School of Hygiene & Tropical Medicine, London;

[2] *United Planters Association of South India,*
Nilgiris, Tamil Nadu, South India;

[3] *Community Health Cell, Bangalore, India*

INTRODUCTION

The papers given so far in this meeting have reported on studies of many aspects of work performance, affecting different occupational groups with varied employment and patterns of work. Each gives a picture of one small part of the complex life of an individual or of an occupationally-defined group. This paper looks ahead to research needs and opportunities, and specifically to a situation where occupational performance can be related to the other aspects of life and health.

The conference has brought together workers of many disciplines, it has shown the relation of nutrition, of physical environment and of infection to work performance and productivity, and by doing this has shown in a precise and quantitative form what had been vaguely and generally felt, that productivity depends on the worker's wellbeing in a set of complex ways. Such a conclusion has several implications for the pattern of future research: it shows that quantitative studies are possible and opens up the linkages between health broadly defined and the economy. It suggests also that the demarcation between work and the rest of life, so convenient for legislation and for the development of the profession of occupational health, has marked limitations. But if we are to study productivity and occupational performance in the context of the rest of life there are real difficulties in finding a tractable system. This paper will show that plantation agriculture is particularly suited for such work, meets a number of other needs in relation to occupational health research, and is a neglected but very promising field for work. It may serve also to stimulate the general discussion which follows, by setting the types of work discussed so far into a broader context.

The particular form of plantation agriculture for discussion is the tea estate of South India.   The paper sets out the characteristics of such an estate, provides a brief description, and describes some very simple and preliminary studies which illustrate its suitability as a research setting.

Within occupational health, some occupations and health risks have been exhaustively studied while others have been largely neglected.   Radiation hazards and pneumoconiosis are the subject of intensive study, as are lead toxicity, and fatigue in aircrew.   Some other trades and hazards are ignored. The largest occupation in terms of workers involved, which has been largely neglected, is agriculture.   The majority of farmers and other agricultural workers are third world nationals, and even less research has been done on the health of third world agricultural workers than of those elsewhere.   There are several reasons: the general paucity of research on health problems of the third world, problems of working with subsistence agriculture of an unorganised type, and lack of facilities for the few research workers interested.   Plantation workers provide one of the better opportunities for work of this sort.

Most occupational research has been directed to populations in which work is clearly demarcated from other aspects of life, and where specific occupational risks from otherwise rare toxic substances or dangerous procedures have been amenable to analysis.   But in the agricultural sector, apart from pesticide toxicity, the distinction between life and work is less sharp.   Often, the two are wholly intertwined.   This provides the research worker with practical difficulties.   Some parts of the life pattern may be inaccessible to study and the holistic approach that the situation seems to need fits badly with the analytical approach.   One needs a situation where most of the worker's life cycle is accessible, where the migratory track is confined, and where it is possible, even if with difficulty, to measure most of the relevant variables. The tea plantation provides such an environment as will be described.

If the control, and even more the eradication, of a communicable disease is to be attempted, an isolated area is desirable.   The association between islands and eradication campaigns for malaria is not accidental.   For most epidemiological purposes a closed population is a great convenience, and for assessing the control of communicable infections in a limited area the closed system is of particular importance.

CHARACTERISTICS OF TEA ESTATES

The key features of a tea estate which make it such a suitable community for research fall into three main categories: (i) it forms a comprehensive community, all-inclusive in many respects;   (ii) it is relatively closed and

Table 1: Some characteristics of tea estates

Comprehensive community
    Both workplace and ecosystem
    Total habitat
    Single health care provider
    Most food acquired on estate

Isolation
    Very low turnover of labour force
    Low migration across boundaries
    Often ethnic minority
    Limited interaction with population outside

Relevant variables measured
    Good demographic data
    Productivity, especially of women, measured daily
    Climatic data recorded

Other features
    Homogeneous group
    Highly organised community

isolated; (iii) several aspects of work and environment are regularly measured (Table 1). The tea plantation provides a defined population in a defined total environment. Almost everything affecting the lives of the tea garden labourer and their families takes place within the plantation. The population is very stable, for the most part consisting of transplanted ethnic groups who have lived for two or three generations on the plantations, and permanent migration into or out of the plantation is very low at 2% annually. Housing, with the related water supply and sanitation facilities, is provided on the plantation, usually dispersed into several small settlements, and health care is provided from a dispensary or a garden hospital on site. Food is obtainable from a shop on the estate, and except where an estate is sited near a town most of the inhabitants of the estate will be employed on site or not at all. The population thus resembles that of a small island in many respects, and flows of people, energy, nutrients, and funds are small across the estate boundaries, and some are measurable. While their parents are working, children below school age are left at nurseries or creches on the estate.

In even moderately organised estates, good demographic data are available because the system of employment, allowances, and maternity benefit require accurate data on the population and it is on balance in the interest of all parties concerned to make such information available. Where health care is under some central supervision or control as is provided through the United Planters' Association (UPASI) in South India and through the Ross Institute's influence more widely, mortality and illness returns are available and may be relatively accurate.

The work is well-defined, and the largest single operation, the plucking of the tea leaves, is accurately monitored by weighing the amount of tea plucked daily by each worker.   This is because of the payment system, which includes productivity bonuses.   Therefore productivity studies are completely feasible. The combination of these data with information on life style that is accessible because of the nature of the estate community makes possible studies of the type set out in the latter part of this paper.

The scale of plantation agriculture is considerable.   In the case of tea alone, it has been the largest earner of foreign exchange in India.   That country grows approximately one third of the world's tea and over 800,000 workers are employed in the tea industry which is the subject of comprehensive labour welfare legislation.   In South India the Planters' Association initiated a comprehensive labour welfare scheme to improve health and living standards, covering maternal and child health, nutrition, environmental sanitation, family planning, child development, and education on personal hygiene and the planning of leisure (Rahmathullah, 1982).   Working conditions and welfare of workers in tea plantations are much better than of others in rural areas, and several of the work tasks on the estate, including plucking the tea, are highly skilled.

The type of study this situation makes possible includes the effect not only of illness on productivity, whether measured in time off work or yield changes during the working days, but also of illness in other members of the family: the effect of childhood illnesses on maternal productivity can be determined in a far more direct way than is usually feasible.

It is also possible to assess the consequences of public health interventions in terms of productivity of the family, so that if water supplies are improved, and this has an effect upon diarrhoeal and cutaneous diseases, the consequences can be traced through the family to measures of economic activity.   It is of course true that this may not disclose all or even the majority of the consequences of diseases or of its prevention, and that a plantation economy is a somewhat atypical one, but nevertheless it is possible to come much closer to measuring productivity as affected by health than is usually the case.   Also, the measurement of production does not require a major innovation as the weight plucked is measured in any case and the only research modification may be in terms of more accurate scales and quality control of the weighing.

The tea estate is not only an industry but is also a community. It comprises a group of people who live in the same area and interact socially. It is an unusual community and one highly suitable for health and health services research. It is a highly stable community with an immigration/emigration rate of approximately 2% per year. Except on estates near to towns, there is a single health service and all on the estate who are ill tend to use it. Alternatives are far away and cost money unless the patient is referred by the estate health services. It is believed that alternative traditional health services are relatively unimportant although this has not been adequately investigated. A closed community with a single health care service is the epidemiologist's dream for many purposes: when in addition there are daily productivity data, in terms of tea plucked, for many of the female population and often climatic data as well, the opportunities for study are apparent.

Although the population of a given estate will be relatively homogeneous, there are marked differences between estates in welfare and in the provision of effective health care. Because services are under the estate's control, the possibility of substantial change is real, and studies of the effect of health care interventions upon morbidity are feasible.

THE POPULATION AND ENVIRONMENT OF A TEA ESTATE

Tea plantations are diverse in size and location. In South India most estates are sited on hilly ground and over 1,000 metres in altitude where rainfall is high, and temperature low. It may be windy.

A middle-sized tea estate, sited 12 km from Coonor in the Nilgiri hills of South India, studied by Narayan and Ramachandran (1982), in a collaborative study between the Ross Institute Unit at St. John's Medical College and the Regional Occupational Health Unit of the Indian Council for Medical Research, had 157 hectares (1.57 km^2) planted with tea. The total population was 1189 with equal numbers of either sex and 35% of the population aged under 15 years. Seventy-six percent were Hindus, and the remainder Christians (Table 2). Thirty-six percent of the total adult population were illiterate, but of those aged 15-19 years only 13% were illiterate and 78% had education beyond primary level.

The age and sex distribution of the population is shown in Figure 1; both fertility and mortality were much lower than in neighbouring non-plantation areas. The 1189 people comprised 261 families, and 76% of those aged between 18 and 60 years were employed, largely on the estate. This particular plantation was not far from a large town so that a larger proportion (ca. 6%) were employed off the estate than is usual.

282 · D. J. Bradley et al.

Table 2: Population of a medium-sized tea estate in Coonor, Nilgiri Hills, South India

A. Total population — 1189
- Males — 598 (50%)
- Females — 591 (50%)
- Hindu — 903 (76%)
- Christians — 286 (24%)

B. Working population — 458
- Males — 271
- Females — 187
- Factory workers — 30
- Field workers — 403
- Skilled workers — 13
- Administrative staff — 12

C. Health Care Team
- Medical Officer — 1
- Compounder — 1
- Nurse — 1
- Midwife — 1

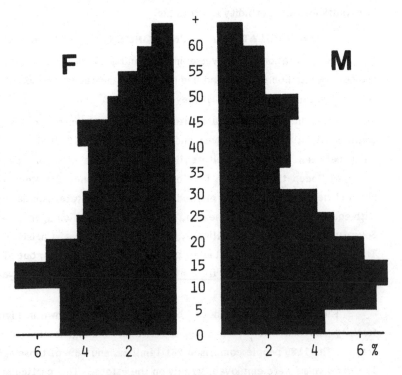

Figure 1: Age and sex distribution of the total population of a tea estate in Coonor, South India.

The families lived in eight clusters of labour lines spread over the estate, except for the 13 management and administrative staff who lived in scattered bungalows and here, exceptionally, there were 41 families in an adjacent village.

PRODUCTIVITY STUDIES ON SOUTH INDIAN TEA PLANTATIONS

The last part of this paper illustrates the points raised in terms of pilot studies already carried out using very simple questionnaires, measurements and observations first to look at the health-related determinants of productivity and then at the consequences of a controlled medical intervention.

The conceptually very simple and preliminary studies that have been carried out so far on a plantation of the type described will serve to illustrate the opportunities outlined above. In 1971 the planters in South India established a department to look exclusively at the health and welfare of labour. As its programme developed, interest grew in the productivity of labour. This was a natural development of the health and welfare programme as one of its main aims was to persuade the employers to accept that improved health and welfare for labour was not charity but could be correctly viewed as development of human resources which comprised the major asset of the planters. A major thrust of the programme was therefore to determine the factors that influenced work performance so as to be able to institute interventions and programmes to provide an environment for better work performance. This would entail better health for the workers and their dependents, improved living environments and meeting basic needs such as housing, water supply, sanitation, medical care and health education.

On reviewing the literature on South Indian plantations and their productivity it became clear that there was a substantial literature on the productivity of tea, coffee and other bushes and no information on the productivity of labour. This clearly required investigation. The tea plantations have the following suitable features:

i.   56% of the employees are women; productivity can be easily measured in terms of the quantity of green leaves plucked. Moreover, the records of weight of tea plucked per person are well maintained on plantations as although basic wages are paid, each woman is also entitled to an incentive wage when extra green leaves are plucked.

ii.   Plucking of green leaves, which is a major task on tea plantations, is skilled work and is done by women in South India. The period of training required to reach optimal skill in plucking is four years. It is an outdoor operation carried out standing between the rows of bushes which come up to the chest level of the workers.

iii.   The women are alloted a number of rows of bushes which need to be plucked. The level and shape of the bushes has to be maintained and the plucking has to be selective to include only the top two leaves and a bud. Thus the worker moves up and down the slopes in between the bushes. The plucking is done by both hands using the thumb and forefinger. The leaf and bud plucked are held in the palm until the palm is full and then they are thrown into a basket suspended on the back by a cloth strap from the head. A high degree of coordination of the eye and hand is developed. The worker carries the weight of the basket and the leaves in it as she continues her work and moves up and down. She is under constant supervision. The weather, though cool throughout the year, is wet for six months of the year.

DETERMINANTS OF PRODUCTIVITY

The first study done was to estimate the existing pattern of productivity among female labour (Rahmathullah & Pothi, 1981). For this purpose a typical medium-sized estate situated at a height of 1965 metres above sea level was selected. The estate of 400 acres had 2700 bushes per acre with an annual yield of 2300 kg of made tea or 9.2 tonnes of green leaves per acre. The estate has gentle slopes but lacks steep gradients. Two hundred and forty women were employed. They were rotated between the fields and could be considered to have a homogeneous working environment. The productivity data for each woman over one year on weight of leaves plucked was tabulated on both a daily and a monthly basis.

The maximum number of days that could have been worked in the year was 291, but no worker was present on every day. After a break from working the daily performance rose by 4.4 kg from the mean whilst the lowest average yield was on pay day.

Age had a complex effect on plucking. Over the age of 45 years the yield was low. Below that age, the women aged 20–30 years had a higher mean yield per day than those over 30 years of age, but when the monthly figures were considered the younger women had plucked less than those over 30 years old because they worked for fewer days. After the first four years of plucking, experience did not seem to increase yield further.

It was found possible to classify the women into three broad categories: consistently good pluckers who had a mean yield of 20.3 kg per day, consistently poor pluckers with a mean yield of 14 kg per day, and 42% of erratic pluckers whose yield was extremely variable between days. The record performance observed that year was by an erratic plucker who brought in 102 kg of green leaves on a single day.

This needs to be viewed in terms of the skill and speed involved. The mean weight of the two leaves and bud, the unit plucked, gives 2000 units per kilogram of green leaves. On this basis, a good plucker producing 20.3 kg in an 8-hour working day is plucking 1.4 units per second, using both hands at once.

RELATION OF ANAEMIA TO PRODUCTIVITY

A pilot intervention study was then undertaken on the effect of anaemia on productivity. Two hundred and forty workers were available on the estate, of whom 12 were excluded from the study due to pregnancy, chronic illness, or incipient retirement prior to completion of the study. The women were randomly allocated between four groups of which A1 and A2 received iron tablets with 65 mg of elemental iron daily, and B1 and B2 received placebo tablets of similar appearance. Groups A2 and B1 received a single course of the anthelminthic Alcopar. The iron trial was on a double-blind basis.

The outcome measures studied before and 100 days after beginning the interventions for all groups were: weight of the worker, haemoglobin level, mean days worked per month over a six-month period, and mean plucking performance in kg of leaves per month. The haemoglobin levels had to be estimated by the unreliable Sahli method as the only one available on the estates at that time, but variance was minimised by the use of the same technician throughout. But the values obtained should only be viewed as comparative indices of haemoglobin level, and the study as indicative of what may be done rather than as a definitive analysis.

The results are given in Tables 3 and 4. There was a rise in haemoglobin level of +1.7 G/100 ml and in weight of +2.6 kg among those receiving iron and a non-significant fall in these measures among the placebo group. The group receiving only iron supplements showed a rise of two days a month in the time worked and an increased yield of 92 kg/month, a 22% rise in productivity, while over the same period the two groups not receiving iron showed a 45.5 kg/month fall in yield, probably because of the difference in season as compared with the baseline observations. Those who received worm treatment and iron showed a small rise in yield, which is not fully explained, though their change in yield was much more favourable than for the two groups not receiving iron.

The study was followed by an open-ended interview of management, supervisory staff and workers. The management and supervisors appreciated the

Table 3:   The consequences of providing iron supplements and anthelminthic chemotherapy to women tea pluckers in South India.   Effects on haemoglobin levels and body mass.

		Before		After 100 days	
		Hb g/dl**	Weight kg	Hb g/dl**	Weight kg
A1	Iron only (56)*	6.2 G	47.8	8.6 G	51.3
A2	Worm treatment with iron (55)*	6.1 G	48.5	6.8 G	51.3
B1	Control (59)*	5.9 G	46.7	5.9 G	45.8
B2	Worm treatment only (58)*	6.0 G	47.7	5.8 G	46.8

*   Number of women in the group
**   Relative haemoglobin level: the method used cannot be depended on for absolute values.

Table 4:   The consequences of providing iron supplements and anthelminthic treatment on the productivity of women tea pluckers in South India.

		Pre-project (3 months)	Project (100 days)	Post-Project (3 months)
		Weight of leaves plucked per month (kg)		
B1	Control	352	302	311
B2	Anthelminthic only	392	342	350
A1	Iron only	409	472	530
A2	Anthelminthic + iron	368	377	374
		Mean number of days worked per month		
B1	Control	20.0	18.0	17.9
B2	Anthelminthic only	20.0	20.5	20.6
A1	Iron only	21.0	22.5	23.5
A2	Anthelminthic + iron	20.0	21.1	20.6

rise in yield but commented most favourably on the increase in number of days worked and on an improvement of personal relations in the field.   The female workers strongly emphasised that those who had received iron had a sense of wellbeing that was absent from those in group B.

EFFECTS OF PREGNANCY

As maternity is a fact of life among female workers, it was decided to study the plucking performance of female workers during pregnancy and one year after they rejoin duty on completion of maternity leave.

The women were matched for age and parity for four groups of women by age and each group had twenty women.

It was observed that during pregnancy, productivity was not affected until the women went on maternity leave 6-8 weeks before delivery. Women took, on average, 100 days for maternity leave as against an authorised 82 days. Primiparous women took more leave than women in the subsequent parities. They attained their average plucking performances in the 5th month after returning to plucking. Then the plucking performance went down from the 6th month to the 8th month and picked up in the 10th and 11th month. They maintained their averages once again in the 12th month.

DETERMINANTS OF PRODUCTIVITY AND WORK ATTENDANCE

A study to assess common factors among 100 persistently poor pluckers revealed that several factors were disproportionately common. Many come from extended families where the occupants of the standard house were more than five in number and included adults other than the husband and wife. Some women had chronically ill husbands or children. The family debt was more than 1500 rupees. The women had an educated unemployed youth in the family.

A study on absenteeism among 680 women and men workers showed that 80% of the women workers stayed away from work due to feeling tired and only 20% of women took leave for social causes. There was no age trend in absenteeism among workers.

Table 5: Symptoms and other findings with a significant (p <0.01) occupational excess for field workers on a tea plantation as compared with those working neither in the field nor the tea factory, on the same estate.

	Prevalence %		Excess prevalence % in field workers
	Field	Indoor	
Joint pains and backache	47	28	+19
Pallor	41	16	+25
Faecal helminth eggs present	18	5	+13
Leech bites, history	77	33	+44
Insect stings, history	24	8	+16
History of peripheral neuritis	9	1	+ 8

The young male workers, aged 20-24, were more often absent than men over the age of 40. The young men took leave for the following reasons: to visit town; to negotiate loans from banks; to look after their land and to cultivate it during the sowing and harvesting seasons when wages are high.

OCCUPATIONAL DISORDERS

In a community-based study, when the frequency of signs and present and past symptoms among those who worked in the fields was compared with the frequency in controls from the estate who worked neither in the fields nor in the factory, five symptom groups and one laboratory finding showed an excess prevalence rate in the field workers, significant at the 1% level (Table 5). Attacks by leeches and insects were higher, as might be expected, as were joint pains and pallor.

CONCLUSION

These preliminary studies show the potential of the tea estate to provide research questions in rural occupational health and also the opportunities to answer them. The relatively homogeneous, comprehensive 'island' community with quantitative evaluation of individual productivity on a daily basis and a health care system on the estates provides a particularly favourable situation for applied research.

REFERENCES

Narayan, R. & Ramachandran, C.R. (1982). Health status of tea plantation workers with special reference to their occupation. Joint Study Report of the Ross Institute Unit of Occupational Health, Bangalore and the Regional Health Unit (ICMR), Bangalore.

Rahmathullah, L. & Pothi, K.S. (1981). Productivity of women pluckers in a typical South Indian tea plantation. Indian Journal of Industrial Medicine, **27**, 447.

Rahmathullah, V. (1982). Comprehensive labour welfare scheme in South Indian plantations. Tropical Doctor, **12**, 71-72.

INDEX

Aborigines, Australian 17, 205
Absenteeism 227, 285
Acclimatisation 8, 206, 208
Acculturation 11

Activity level 59, 76, 146, 150-154, 193,
  215, 217, 223
    daily 3, 161, 198, 221, 222, 250
    pattern 16, 18, 19, 41, 54, 160, 181,
      197, 217, 221, 223, 258, 266

Adaptability 6, 20, 71
Adaptation 2, 15, 71, 77-80, 82, 87, 96,
  100, 130, 132, 147-8, 156, 165, 168,
  181, 206, 219

Adaptive mechanism 145
    response 62, 87
    strategy 143, 145, 154

Adipose tissue, adiposity 79, 109
Admixture (intermixture) genetic 86, 90
Adolescence 60
Adulthood 62, **107-140**

Aerobic capacity 31-34, 46, **51-56**, 63,
  95-97, 126, 127, 182, 209, 239
    power 4, 7, 8, 11, 13, 16, 182, 185
    work 85, 95, 181, 209

Africa(n) 1, 3, 4, 7, 10-14, 17, 18, 37-39,
  42-45, 117, 158, **165-179**, 193-203,
  227-234, 235, 247, 248
Age(ing)(s) 34-39, 42, 43, 46, 54, 59, 90,
  91, 94-96, 100, 117, 120-124, 127,
  132, 134, 149, 167, 171, 181, 185,
  193, 194, 197-199, 215, 217, 223-225,
  227, 231, 236, 240, 268, 279, 282, 285
Agility 181, 186

Agriculture(al) 1, 158, 193, 196, 198, 199,
  221, 248, 257, 267
    activity 3, 156, 157, 199, 253, 258,
      266
    workers 66, 110, 221, 236, 237

Alpaca herding 259
Altiplano 252, 253, 256, 257, 264, 267, 270

Altitude 33, **85-106**, 185, 217, 220
American College of Sports Medicine 20
Amerindians 85, 86, 90
Amoebic infection 10
Anaemia 10, 19, 59, 61, 65, 67, 68, **85-106**,
  200, 227, 244, 283

Anaerobic capacity 3, 4, 7
    power 3, 4, 7
    threshold 8, 9, 14

Ande(s)(an) **85-106**, 252
Angina 52
Animal production 199, 263
Ankylostomiasis 198
Anthropometr(y)(ics)40, 80, 121-123, 220,
  259, 262, 263, 267, 268
Antigens 227, 244
Antihelminthic 283, 284
Antilles 256
Arable land 223, 224
Arms 117, 119, 120, 262
Ashkenazi 207, 209
Asthma 237
Astrand and Rhyming nomogram 193-5,
  197-9, 200
Atherosclerosis 51
Athletes 5, 31, 40, 52, 65, 228
Atmospheric pressure 85
ATP 114

Bacterial infection 227
Bantu 4, 14, 17, 34, 235
Basal metabolism (BMR) 80-82, 130, 141,
  146, 147-154, 156
Basketball 5
Bedouins 207, 209
Bed-rest 34

Behaviour(al) 218
    adaptation 149, 161
    mechanism 148
    patterns 18, 156

Bicarbonate 101
Biceps 117

Bicycle ergometer(ry) 79, 96, 120, 131, 132, 197-200, 209, 211, 218, 220, 228, 237, 239, 268
Biological impairment 251, 252

Blood films 237, 241
    pressure 14, 219, 220
    volume 12, 79, 101, 112, 168, 229, 231

Body build 10
    compartments 112, 113
    composition 6, 11, 12, 65, 78, 80, 81, 110, 112, 121, 125, 168, 169, 177, 182, 211
    dimensions 141, 145, 146, 228
    fat 3, 5, 6, 11, 12, 54, 63, 82, 109, 117, 118, 169, 171, 211, 220
    mass 3-7, 10-12, 35, 36, 52, 53, 55, 58, 62, 78, 80, 109, 114, 117, 120, 126, 133, 145, 146, 169, 171, 197, 201, 209, 211, 236, 239
    miniaturisation 141, 148
    muscle 32, 53, 239
    potassium 53
    proportions 60, 168, 169, 170
    reserves 78
    shape 40
    size 7, 10, 12, 15, 35, 36, 46, 59, 60-64, 78, 119, 128, 135, 145, 147, 148, 154, 168, 169, 171, 172, 177, 181, 182, 185, 189
    water 5, 126
    weight 35, 81, 82, 96, 109, 110, 112, 114, 117, 120, 124, 130, 133, 134, 145, 147, 155-156, 159, 172, 196, 207, 210, 211, 228, 229, 236, 239, 268

Bolivia **85-106**

Bone cross section 168
    diameters 83
    marrow 94
    mineral 5, 6, 110

Brain 61

Breast feeding 250
    milk 59

Breathlessness 51
Building industry 115
Bushmen 4, 14, 17, 18
Byssinosis 42, 43

Calf 117, 209-211, 213
Caloric intake, calories 65, 66, 107, 110, 120, 130
Calories 65, 66
Cameroun 193, 194
Canal cleaners 238, 239

Cane-cutt(ers)(ing) 9, 14, 33, 63, 110, 115-119, 227, 235
    loaders 116, 119

Carbohydrate 110, 116
Carbon monoxide 100

Cardiac dilatation 228
    frequency 31, 32, 35, 37, 53, 55, 95, 98
    muscle 112
    output 15, 101

Cardiovascular adaptation 35
Caribbean 37, 41
Caste(s) 215, 216, 218, 219, 221, 223, 224
Catch-up growth 145

Cell function 110
    mass 53, 114, 239

Chemotherapy 236, 245
Chest 31, 40, 42, 117, 262
Child care 250
Childhood 40, 58, 61, 107-140, 261, 278
Children 11, 36, 40-41, 45, 52, 59, 61, 62, 71, 88, 90, 143, 165-179, 228, 230, 277
China (Chinese) 11, 36, 242

Cigarette smoking 4, 45
Circulation 31, 244

Circumference, arm 63, 117, 119, 262
    calf 63, 117
    chest 117, 262
    limb 82, 117, 118
    thigh 209

Clerks 35
Climate 16, 83, 206

Coal industry 43
    miners 33, 36, 42, 44, 45, 65, 66

Cognitive development 61
    performance 61
    skills 71

Colombia 12, 14, 17, 63, 108, 117, 120
Commercialisation 247
Compensatory responses 256
Construction workers 221
Cooking 78, 82

Coordination motor 169, 171
    hand eye 282
    manual 186

Copper-miners 40
Cost-benefit analysis 2
Crafts(men) 221
Creatinine excretion 109, 110, 112
Crops 2, 263
Crops, cash 2, 3, 17
Cultural change 20
Cyclic variations 16

Daily exercise 16
Debility 235
Decade of the Tropics 1, 19, 20, 46, 55
Deforestation 247, 249
Degradation, of environment 247, 249
Dehydration 11
Demograph(ic)(y) 217, 247, 249, 261
Desert 1, 206, 249

Development 58, 166, 278
-al effects 58

Diabetes 81, 237
Diarrhoea 59, 237
Diet 15, 59, 62, 78, 81, 110, 116, 166, 197

Dietary intakes 59, 66, 143, 160
repletion 110, 111, 113-115
supplementation 11, 119, 155
surveys 16, 81, 82, 160

Digestibility 17

Disease 2, 247, 249, 250, 253, 254, 257
response 141
vectors 217

Ditch digging 64, 119, 237, 239
Divers 52
Douglas Bag 82, 236, 237, 268
Drought 141, 247, 269
Dust 34, 42, 43, 46
Dynamometers 6

Earnings 71

Economic activity 79, 218, 221, 223
development 57, 72
status 216

Ecosystems 2
Education 2, 72, 278
Efficiency 51, 70, 130-132, 134, 143, 172
Egg excretion 238-9, 242, 245
Egg loads 237, 242-245
Elderly 4, 6, 25, 226
Electrocardiograph(y)(ic) 52, 207, 209, 229
Electroencephalography 186
Electromyography 211
Employment 2, 6, 63, 224, 277, 279
Endurance 8, 11, 109, 110, 114-116, 181, 182, 228, 250, 267
Energy balance 69, 70, 141, 143
content 16, 78, 81
cost 2, 63, 82, 145, 147, 149, 159, 160
deficit 57, 70
equilibrium 78, 141-143, 148, 149, 156
exchange 141, 205, 206
expenditure 10, 16, 17, 66, 69-71, 78, 79, 82, 133, 146, 147-153, 156, 159-161, 172, 177, 198, **205-213**, **235-246**
intake 18, 59, 61, 65-69, 71, 78, 79, 110, 141, 148, 154, 155, 159, 172, 177
malnutrition 141-164
needs **57-75**
nutritional status 141
output 141, 147, 149
requirement 69, 71, 78, 141, 145, 146, 177
reserves 156, 177
saving 145-147
sparing **141-164**
stores 69, 71, 141, 142
supplements 65, 66, 69
transfer 15
utilisation 17, 69, 70, 143, 147
yield 17

Enzymes, glycolytic 114
oxidative 114

Epidemics 258, 269
Ergometer 7-9, 13, 35, 51, 96, 114, 193, 195, 245
Erythrocyte protoporphyrin 94
Eskimo 8, 205
Ethiopia(ns) 4, 14, 80, 120

Ethnic differences 33, 35-37, 39-41, 60, 94, 194, 195, 206, 209
groups 15, 19, 35, 37-39, 41, 60, 143, 145, 169, 171, 207-209, 211, 216, 218, 277

Ethnicity 36, 46, 182
Europe(ans) 37-42
Excavation 119

Exercise 11, 15, 16, 51, 100, 103, 116, 120, 172, 182, 195, 199, 201, 205, 217, 219, 228, 230, 235, 268
capacity 52, 83
routine 149
tests 20, 199, 200
physiology 6, 215

Exertion 19
Exhaustion 114
Explosive strength 6
Extracellular fluid 126

Factory work(ers) 31, 35, 64, 119

Family 224, 266
care 250
planning 278
studies 33, 182, 183, 185, 186, 189, 259

Famines 141

Farm labourers 65, 95
Farmers 35, 198, 213, 218, 228, 230, 231, 236, 239, 242, 247-274, 276, 278

Fasting(s) 16, 18, 83, 147
Fatigue 15, 51, 235, 276
Fat-free mass 35, 36, 52-54, 147
Fertility 279
Fetus 79
Fevers 59
Filariasis 10, 198
Firewood 250
Fishermen, fishing 96, 108, 109
Fitness 11, 13, 16, 62, 165-179, 181, 182, 217, 221, 222, 235
Flow indices 43

Fluid balance 12
loss 11

Food 78, 81, 269
availability 71, 77, 79, 80, 143, 148
energy 57, 78, 154
intake 16, 19, 80, 81, 177
producers 2
production 247, 248
resources 156
shortage 141
supply 78, 80

Forced expiratory volume (FEV$_1$) 31, 35-
    37, 43, 183, 185, 220
  vital capacity (FVC) 31-34, 37-40, 42,
    43, 45, 53, 183, 185, 220

Functional capacity 250, 251, 255
  impairment 254

Gambia 71, 80, 155, 161
Gas exchange 9
Gatherers 16

Genetic(s) 33, 181-192, 205, 206, 210
  diversity 34
  endowment 16, 127
  isolates 207
  potential 60, 143, 146, 165, 177
  variance 15

Geographic location 211
Germany 65
Gezira 236-244
Glucose-6-phosphate dehydrogenase
  deficiency 207
Glycogen 11, 116
GNP (Gross National Product) 143
Grip strength 269

Growth 10, 36, 41, 59, 60, 62, 107, 118,
    122, 125, 141, 143, 145, 146, 165-179,
    263
  failure 108
  potential 143
  retardation 62, 135, 143, 145, 167,
    177
  velocity 122, 123, 125

Guatemala 11, 17, 63, 66-69, 109
Guyana 38, 45
Gymnasts 228

Habitual activity 16, 54, 189, 206, 217
Haematocrit 110, 228, 231, 233
Haemoglobin 31, 33, 87-89, 91-93, 95,
    100, 110, 112, 119, 228, 230-233, 240,
    241
  concentration level 13, 67, 87, 88, 90,
    94-97, 100, 114, 201, 202, 215,
    229-232, 236, 239, 244, 245, 283,
    284
  distribution 89-92, 95, 103

Haemopoiesis 87, 94, 95
Hand steadiness 186
Handgrip 4, 12, 13, 171, 174, 186, 187
Harvard step test 67
Harvest(ing) 77, 83, 181, 250, 286

Health 2, 33, 46, 69, 78, 86, 94, 181, 217,
    218, 250, 252-254, 256, 257, 259, 261,
    262, 275-277, 281
  care 277, 279
  problems 85
  status 206, 220, 254, 262, 269
  symptomatology 262

Heart disease 237
  rate 8, 9, 13, 14, 18, 35, 62, 96, 100,
    120, 193, 194, 196, 201, 207, 209,
    217, 219, 220-223, 268
  volume 168, 228, 229

Heat, environmental 12-15, 52, 177
  loads 206, 208
  loss 10, 205

Heavy labour 221, 222, 265
Height 3, 5, 10, 12, 39, 42, 59-65, 82, 85,
    96, 100, 110, 111, 117, 119-123, 130,
    143, 145-146, 165-168, 201, 209, 210,
    229, 231, 240
Herder(s)(ing) 218, 252, 256, 259
Heritability(ies) 15, 182, 185, 186, 188,
    189
High altitude **85-106**, 217
Highland(ers) 9, 35, 41, 206, 211
Hillsides 2, 36, 63
Hindu 216, 218-226
Histopathology 227
Hookworm 10, 232, 235, 237

Household 217, 223, 224, 251, 256-258,
    261, 263, 266, 267
  tasks 159

Housework 82

Housing 281
  conditions 143, 262

Humidity 13, 52, 165, 177, 220
Hunting(ers) 3, 16, 181
Hybridisation, hybrids 10, 189
Hycanthone 242, 244
Hydration 11, 12
Hygiene 166, 278
Hypercapnia 183, 185
Hypertension 237
Hyperventilation 100
Hypoxia 85, 87, 95, 103, 183, 185

Illness 71, 247-274, 278, 283
Immigration 206
Income 59, 216
India(ns) 3, 5, 6, 10, 11, 17, 18, 35, 37-39,
    41, 42, 52, 59, 60, 62, 64, 66, 80, 147,
    276, 278, 281
Indonesia 18, 67, 130, 147

Industry 279
  cottage 2, 69
  labour intensive 2

Infancy, infants 59, 71
Infection(s) 59, 62, 71, 243, 275
Influenza 250, 269
Intelligence 58, 61
Intermittent claudication 52
International Biological Programme 1, 3-
    5, 7, 15, 16, 19, 31, 37, 40, 53, 193
International Council of Sports Medicine
    20
International Labour Organisation 20

Iron 66, 67, 283, 284
  binding capacity 94
  deficiency 68, 87, 89, 91-95, 103
  therapy 13

Irrigation 237
Islands 2
Isolation 207
Israel 18, 206-213

Italy 143, 158, 159
Ivory Coast 159

Jamaica 4, 17, 63
Japan(ese) 61
Java 18, 130, 147
Jerusalem 206, 209-211, 213
Jumping 57, 65, 171, 186
Jordan 206

Kenya 64, 66, 119
Knee extension force 4
Kurd(s)(ish) 14, 207-211, 213

Lactate 7-9, 15, 101
Lactation, lactating women 78-81, 155
Latex workers 67, 68
Latin America(n) 17, 65, 89, 91, 108, 247
Lead toxicity 276
Lean body mass 3, 6, 35, 58, 62, 63, 209,
   236, 239, 240, 241, 243, 244
Learning 58, 189
Leg mass **205-213**
   volume 7, 13, 35, 63, 117, 120, 211
Leisure 78, 79, 278

Life expectancy 5
   patterns 206
   span 71
   styles 79

Lifting 4, 6, 169

Limb muscles 7
   width 6

Liver disease 81
   fibrosis 227

Livestock 250
Living standards 227
Load-carrying 221, 265, 268
Logging 116
Low altitude, lowland 41, 85-88, 90-94,
   101
Lumberjacks 117

Lung capacity 65, 182
   diffusing capacity 100
   function 185, 235
   tissue 267
   volumes 3, 168

Malaria 10, 198, 227-235, 237, 241, 244,
   276
Malignancy 227

Malnutrition 5, 7, 10, **57-75**, 77, 86, 107-
   140, 156, 167, 172
   protein energy 86

Manual labour(ers) 95, 218, 221, 223
Marasmus 61

Marginal conditions 247
   nutrition 77-106, 107-140, **141-164**,
   171, 172

Maternity 277, 285
Maturation 121, 122, 168, 170, 189

Maximal aerobic power 95, 96, 109, 112,
   113, 117, 119, 120, 124, 127, 184,
   185, 193, 194, 197-200, 202, 235, 236,
   251, 254
   aerobic capacity 85, 97, 239
   cardiac output 112
   exercise 51, 52, 83, 267, 268
   heart rate 8, 52, 112, 113, 185
   oxygen consumption 109
   oxygen uptake 3, 7, 8, 10-14, 31, 34,
      51-55, 58, 62, 171, 172
   work 268
   work capacity 205, 206, 209, 258

Measurements 62, 80-82, 117, 145, 209,
   238
Mechanical efficiency 15, 108, 132, 134
Mechanisation 107, 160
Medical history 238
Medilog 19
Mental development 61
Mestizo 121

Metabolic adaptation 147-148
   efficiency 143, 147, 155
   heat 205, 208
   rate 78, 79, 81, 82

Metabolism 18, 35, 205, 207, 208
Metabolites 6
Mexico 40, 171
Migrants 38, 41, 42
Migration 185, 247, 270, 277, 279
Milk 71

Mineral(s) 6
   content 6

Miners 40, 43-45
Minnesota Study 148
Mitochondria 51
Morbidity 247, 257, 258, 263, 279
Morphology 263
Mortality 71, 171, 247, 257, 258, 278, 279
Motivation 8, 12, 13, 58, 61, 67, 116, 118,
   128, 172, 243

Motor development 170-172
   fitness 172, 177
   performance 167, 170, 172, 173, 176,
      177, 182
   skills 177
   tasks 171, 172, 174, 186, 188, 189

Mountain(s) 39, 193, 247

Muscle(s) 13, 51, 63, 95, 113, 211
   biopsy 6
   cell mass 109
   contraction 114
   development 6, 7
   enzymes 6
   fatigability 114
   fibre 6, 8, 169
   force 11, 169, 228
   function 6, 114, 116, 124, 172
   glycogen 114, 116
   mass 5, 7, 12, 31, 34, 35, 51, 119,
      122, 126, 135, 171, 172, 205, 228

oxidative capacity 126
strength 3, 4, 6, 12, 13, 31, 117, 168, 169, 182, 187, 189
tissue 114, 116, 182, 228
viscosity 15
volume 236

Muscularity 6, 197
Myocarditis 229

Natural selection 181
Negev 15, 206-209, 211
Nepal 14, 215-226
New Guinea(n) 3, 9, 11, 13, 15, 17, 18, 35, 37, 40, 41, 45, 52, 156, 158, 171
Nigeria(ns) 5, 18, 33, 36, 40, 43, 44, 45
Nunoa 253, 257-259, 267-269

Nutrient availability 59
content 81
intakes 59

Nutrients 57, 59, 277

Nutrition(al) 2, 10, 11, 19, 42, 46, 51, 71, 72, 86, 166, 172, 177, 181, 235, 275, 278
conditions 160, 170
deficiencies 57, 112
deprivation 120, 121, 127, 145
inadequacies 250
marginal 77-106, 107-140
status 12, 58, 59, 62, 65, 66, 72, 107, 109, 112, 116, 118, 119, 120, 135, 141, 172, 197, 206, 215, 217, 220, 236, 254, 257, 262, 263, 269
stresses 77, 145
supplementation 59, 67, 114
problems 85

Obesity 95
Occupation(al) 206, 213, 217, 218, 223, 268, 275, 276
Oedema 12
Office workers 2
Onchocerciasis 198, 199, 232
Overnutrition 57
Overwork 250
Oxidation 9, 126

Oxygen affinity 100
consumption 8, 13, 15, 19, 82, 83, 95, 97, 98, 100, 193, 208, 228, 236, 238, 268
costs 15, 19
intake 8, 13, 237, 242
loading 103
pressure 85
pulse 53
tension 85, 99, 100
transfer 31, 32
transport 10, 12, 13, 32, 51, 85, 88, 90, 94, 95, 100, 101, 103, 112, 200, 201, 228, 267
uptake 3, 7, 13, 54, 171, 209, 244

Oxylog 19, 236-242

Parasite 237
smears 233
rate 228, 232

Parasitic disease 227
Parity 285

Peak flow growth velocity 122
flow meter 43
flow rate (PFR) 43, 183, 185

Peasant(s) 2, 3, 158, 215, 225, 252, 256
farmers 147

Pedometers 18
Performance tests 7, 10
Perimyocarditis 227, 228
Peru 252, 253
Pesticide 276
Philippines 17, 67, 80
Phosphagen 7
Phosphocreatine 114

Physical activit(y)(ies) 15, 19, 35, 36, 46, 61, 78-82, 120, 126, 141, 145-149, 155, 158, 171
characteristics 201, 209
condition 119, 127
exertion 252
fitness 52, 172
growth 58, 171
performance 85, 86, 103, 167, 168, 181, 232, 243
training 35, 36, 51, 54
work 51, 109, 116, 127, 129, 135, 146, 148, 171, 232

Physiological adaptation 100
compensation 95
function 245
impairment 256

Physique 40, 182, 189
Placenta 79
Plantation(s) 9, 67, 275, 276

Plasma 101
protein 12
volume 94, 101, 229, 231

Plasticity 156
Play 61
Ploughing 218, 221, 223
Pneumoconiosis 42, 276
Pollution 218
Polycythaemia 90-92, 94-96
Porters 221
Posture 42, 82, 160
Poverty 1, 57
Power output 51-53
Praziquantel 241-243
Pregnancy 58, 59, 78-81, 88-94, 283, 285
Production 251, 256, 263, 269
Productivity 57, 61-68, 71, 103, 107-140, 252-256, 263, 275, 278, 279, 281, 282
Prophylaxis 232, 233

Protein(s) 110, 116, 177
energy malnutrition 86, 171
intake 70, 110, 114, 120

Puberty 122

Pulmonary diffusion 103
  disorders 95
  function 46, 220, 238
  ventilation 31

Pygmy 10, 34, 165, 176, 177
Pyruvate 9

Questionnaires 16, 238, 261-263, 281

Radiation hazards 276
Rainfall 16, 109
Red blood cells 90
Relative humidity 14
Reproductive competence 141

Respiratory disease 258, 264, 267, 268
  failure 40
  function 40, 182, 183, 189, 216, 228
  gas exchange 207-209

Respirometer 19, 82, 236, 237
Reticuloendothelial blockage 94
River blindness 251

Road building 119
  workers 63, 66

Running 5, 7-9, 11, 65, 171
  distances 13
  speed/velocity 169, 186

Samoa(ns) 2, 10, 14
Sampling 9
Sanitation 143, 218, 277, 278, 281
Schistosomiasis 10, 198, 227, 235-246, 256

School achievement 61
  children 93, 120, 121, 202, 235

Season(al) 19, 77, 81, 258, 262, 263, 265,
  267, 270, 283
  body weight 71, 155
  variations 207-209

Secular trend 10

Self-esteem 224
  selection 35

Semistarvation 78, 109, 114, 116, 147, 149

Serum albumin 110, 111
  ferritin 94
  iron 94
  proteins 111

Sex 91, 182, 185, 194, 215, 217, 278
  dimorphism 166, 194, 210
  hormones 169

Sexual maturation 123, 135, 169
Shipyard workers 52, 53
Shivering 205

Shoulder girdle 43
  widths 40

Sierra Leone 65

Singapore 1
Sitting height 40
Skeletal muscle 114, 116
Skill 2, 283
Skin 13

Skinfold 5, 12, 63, 69, 82, 112, 118, 120,
  122, 123, 126, 209, 220
  thickness 4, 169

Smoke 45, 46, 264
Smoking 33, 34, 39, 43, 45, 54

Social behaviour 251, 254, 264
  environment 61, 206
  function 65
  interaction 256
  obligations 250
  organisation 72
  relations 253, 256, 269
  responses 250, 251, 256
  sciences 20
  status 215, 271, 238
  structure 225

Sociocultural influences **215-226**

Socioeconomic conditions 61, 206
  effects 124, 227
  indicators 143
  relationships 247
  status 15, 33, 81, 121, 122, 166, 217,
    257, 262, 264, 265
  strata 143

Soldiers 52, 194
South America(n) **85-106**
Sowing 286
Spirometer 41
Sports 168
Sri Lanka 67
Starvation - see also Semistarvation 1, 11,
  18, 69, 109, 250
Stature 5, 10, 34-40, 42, 43, 62, 64, 86,
  143, 146, 165-179, 182, 197, 235, 239,
  243, 245
Step test 8, 67, 83, 193-197, 200, 237
Strength 2, 3, 171, 172, 177, 181, 186,
  250, 267, 268
Stroke volume 112, 228
Stunting 60, 62, 171
Subcutaneous fat 122, 211

Submaximal effort 228
  exercise 13, 35, 53, 55, 83, 98, 102,
    103, 236, 267
  work 11, 31, 108, 130, 132, 235, 237,
    243, 258, 268
  workloads 96, 112, 219

Sudan 33, 165, 194, 235-246, 256

Sugar plantation 9
  -workers 17, 42, 63, 65, 115

Summer 15, 206-209
Supplement(s)(ation) 11, 66, 67, 156, 283,
  284
Survival 2, 181
Sweating 15, 19, 205
Symptomatology 261, 263

Tachycardia 13, 19
Tanzania(ns) 17, 63, 193

Tea 278, 281
    estate 276, 282
    plantation 275-286

Technology 1, 2, 72

Temperature ambient 15, 51, 205
    body 205
    environmental 3, 14, 18, 52, 165, 177,
        215, 217, 220

Thalassaemia 207
Thermoregulation 165, 177, 205
Thermoregulative capacity 177

Thigh muscle 32
    volume 209-211, 213

Thoracic volume 31, 40, 117
Thorax 31, 42, 46
Throwing 57, 65, 171, 186
Tidal volume 33, 183, 185
Time use 149, 156, 158, 159, 161, 258,
    265, 267, 270

Tissue enzymes 95
    water 5

Tobacco 45, 46
Toxicity 276
Toxins 227, 245
Training 34, 35, 126, 168, 189, 221, 233,
    281
Transfer factor 31
Transferrin 91, 92, 94
Treadmill 51, 67, 83, 114, 115, 132-134
Treatment 236, 242, 244
Trinidad 35, 45
Tropic(s)(al) 10, 85, 206
Trunk 42
Tuberculosis 81, 251, 254
Twins 15, 33, 182, 183, 185, 186, 189

Underdevelopment 145
Undernutrition 45, 57, 107, 110, 120, 130,
    135, 247, 250, 252-254, 269
United Nations 107
Urbanisation 11

Vascular perfusion 100
Vasoconstriction 205
Vasodilatation 295

Velocity height 122
    growth 122, 123
    peak 122
    weight 122

Venezuela(ns) 4, 14

Ventilation 15, 31, 33, 55, 103, 236
    equivalent 98
    rate 236, 238

Ventilatory adaptations 100
    capacity 31-50, 185
    equivalent 53, 100

function 31, 45
response 40, 98, 183, 185
sensitivity 41

Villagers 35
Viral infection 227
Visceral flow 15

Vital capacity 11, 31, 35, 40, 182
    statistics 143

Vitalograph 43
Vitamin C 67

Volume, calf 209, 211, 213
    leg 7, 63, 117, 211
    thigh 209, 211, 213

VO_2 max (see aerobic capacity) 3, 12, 32,
    33, 35, 58, 62, 63

Wages 64, 224, 227, 286

Walking 18, 221, 222, 268
    speed 132, 134

Washing 82

Water 250
    contamination 264
    content 12
    quality 264
    supply 277, 281

Weakness 219
Wealth 224, 265
Weaning 59
Weed-cutting 239

Weight 5, 42, 59, 62-65, 67, 71, 82, 85, 86,
    94, 96, 100, 111, 117, 119, 120-124,
    130, 145, 146, 160, 166-168, 171, 182,
    194, 195, 201, 209, 210, 231, 283
    /height 42, 63-65, 69, 100, 110, 111,
        120, 125, 167

Welfare 278, 279, 281
Well-being 250, 254
Winter 206-209

Work(ing) 45, 51, 236, 257
    capacity 1-30, 31-33, 57-59, 61-63,
        72, 77, 85, 106, 107-140, 141,
        171, 181-192, 193-203, 209-211,
        215-226, 227-234, 250, 253, 254,
        256, 267
    efficiency 121, 130, 171
    effort 220
    habits 238
    load 62, 100, 115, 195, 196, 209, 222,
        223, 250
    muscular 1, 135
    output 57-75, 115, 119, 235, 268
    performance 14, 65, 141, 195, 266,
        270, 275, 281
    potential 252, 254, 256, 258, 267, 269
    rate 9, 63, 237

World Bank 57, 71
World Health Organization 20, 88, 198,
    202, 235

Yemenites 207-210, 213

Zaire 34, 165, 166, 174
Zimbabwe 80
Zygosity 182, 189